Thermodynamics of I.C. Engines

Scott L. Post, Ph.D.

Table of Contents

1.1 HOW ENGINES WORK	2
1.2 OTTO CYCLE	4
1.3 DIESEL CYCLE	9
1.4 DUAL CYCLE	17
1.5 ENGINE CYCLE SIMULATIONS	19
1.6 BRAYTON CYCLE AND GAS TURBINES	31
1.7 STIRLING CYCLE	36
1.8 OTHER CYCLES	39
1.9 FUELS	43
1.10 ADVANCES IN ENGINE TECHNOLOGY	48
1.11 USEFUL WEBSITES	51
THERMODYNAMICS EQUATION SHEET	52
SUMMARY COMPARISON OF ENGINE CYCLES	53
REFERENCES	54
MATLAB CODES	56

The objective of this book is to aid the student in understanding thermodynamics and developing the tools to solve engineering problems involving the application of thermal sciences to understanding and designing internal combustion (I.C.) engines. This book is designed to provide a study aid to students taking an undergraduate thermodynamics or I.C. engines course in engineering.

1.1 How Engines Work

The first law of thermodynamics tells us that energy can be converted between different forms, and Joule long ago discovered the analogy between heat and work. The basic question an engine tries to answer is how to convert the chemical heat of the fuel into useful mechanical work in the most effective way possible. This first sentence feels awkward to me

Nicolaus Otto (1832-1891) developed the first four-stroke internal combustion engine. The four strokes are: intake, compression, expansion (power), and exhaust, as shown in Figure 1.1. In the intake stroke, the piston moves down, creating a vacuum, drawing fresh air into the cylinder. After the piston reaches its lowest point, bottom-dead-center, it moves up during the compression stroke, compressing the air to high pressure and temperature. The fuel-air mixture is ignited near top-dead-center, when the system has its minimum volume. The piston drops again during the power stroke, as the high-pressure burned gases push on the piston top. Then the piston moves up again during the exhaust stroke, which pushes the exhaust gases out of the cylinder so the cycle can begin again.

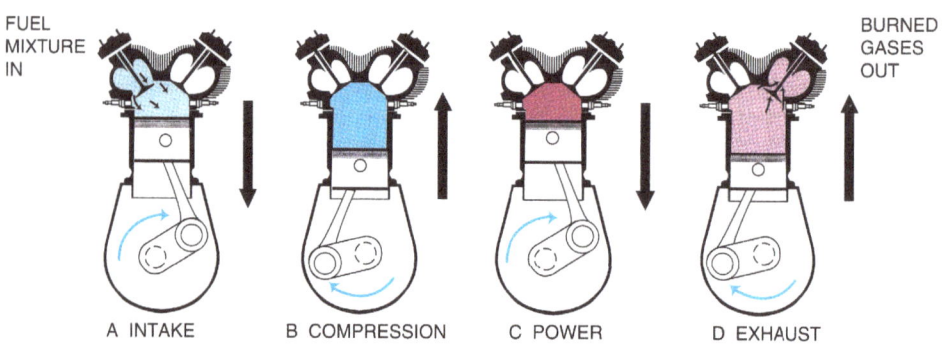

Figure 1.1: Schematic of four-stroke engine. From [FAA].

In almost all modern internal combustion engines, there are valves on the top of the cylinder head for fresh air inflow and exhaust gas outflow. Gasoline engines have sparkplugs on the top of the head as well, while diesel engines, which are compression-ignited, have a high-pressure fuel injector on the head. The port fuel injectors of gasoline engines are low pressure, generally in the (kPa) 40-60 psi range, while the direct-injection diesel engines have injection pressures in the (MPa) 10,000-30,000 psi range. The piston has rings that seal the combustion chamber. An oil film between the rings and cylinder wall reduces friction losses.

All thermodynamic cycles form a complete loop, so that the properties (P, T, V) at the end state are the same as the initial state. The numerical subscripts 1, 2, 3, … are used to denote each state of the system. The cycles that will be discussed in this book are the Otto, Diesel, Dual, Stirling, and Brayton cycles, which are all *gas* power cycles

In all cases we assume the **ideal gas law** holds true, so that the state relationship is:

$$PV = m\frac{R}{M}T \qquad (1.1)$$

Note that in this book for simplicity the symbol R is used to represent the universal gas constant. Also note that for closed systems the mass is constant, so that the ideal gas law becomes:

$$\frac{PV}{T} = \text{constant} \qquad (1.2)$$

If we further assume that the specific heats are constant, then all of theoretical cycle problems can be solved by hand. The specific heat ratio, k, is:

$$k = c_P / c_V \qquad (1.3)$$

The first law of thermodynamics for a closed system is:

$$\Delta E = Q - W \qquad (1.4)$$

Where ΔE is the change of energy of the working fluid (mostly air in the case of engines), Q is the new heat transferred into the system, and W is the net work done by the system. The changes in kinetic and potential energy will be negligible for the air in the engine, so that the change of energy of the air is simple that of the internal energy, so $\Delta E = \Delta U$. For any thermodynamic cycle that completes a closed loop, the fluid returns to its initial thermodynamic state, so the change in internal energy is $\Delta U = 0$. Thus the first law (Eq. 1.4) simplifies to $Q = W$. There is heat added to the system, Q_{IN}, from the combustion of the fuel, and heat exhausted from the system, Q_{OUT}, at the end of the expansion stroke. Thus the next work of the engine can be calculated from the first law energy balance:

$$W = Q_{IN} - Q_{OUT}. \qquad (1.5)$$

The first law efficiency, η, can be calculated based on the energy balance as:

$$\eta = (Q_{IN} - Q_{OUT}) / Q_{IN} \qquad (1.6)$$

Alternatively the work, W, can also be calculated by integrating the boundary work done by the high-pressure cylinder gases on the moving piston over the entire cycle:

$$W = \int P \, dV \qquad (1.7)$$

It is common to assume isentropic compression and expansion processes in ideal engine cycles. An **isentropic** process implies two things: that the process is *reversible* and frictionless, and that it is *adiabatic*. The first assumption is typically better than the second. Due to metallurgical limitations high temperatures cannot be maintained in the engine block, and hence your engines have a radiator and cooling system, which results in heat loss during the power stroke.

1.2 Otto Cycle

There are four processes, going through four state points, which comprise the Otto cycle. The first is isentropic compression from the intake conditions, followed by constant volume heat addition from combustion, then the power stroke, which is modeled as an isentropic expansion, and finally a return to the end state with constant volume heat rejection. These processes are shown on the diagram of Figure 1.2.

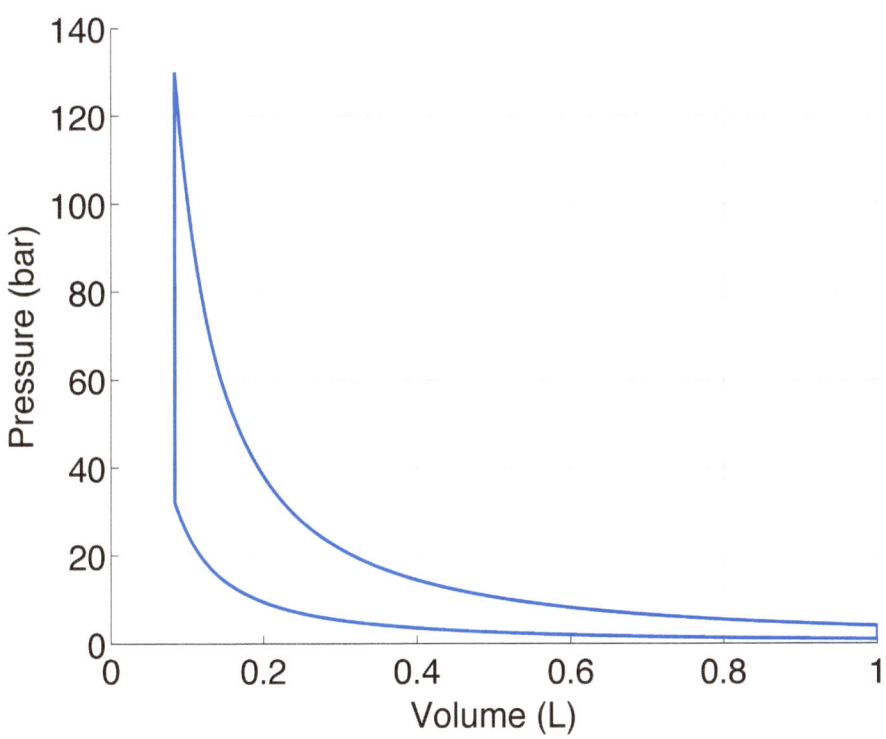

Figure 1.2: Pressure-Volume (P-V) diagram of the Otto cycle, using conditions of Example 1.1.

Table 1.1: Steps in the Otto Cycle

Process	Thermodynamic relations	Volume relations
Isentropic Compression	$P_1 V_1^k = P_2 V_2^k$	$V_2 = V_1$ / Compression ratio
Constant Volume Heat Addition	$Q_{IN} = m\, c_V\, (T_3 - T_2)$	$V_3 = V_2$
Isentropic Expansion	$P_4 V_4^k = P_3 V_3^k$	$V_4 = V_3$ * Compression ratio
Constant Volume Heat Rejection	$Q_{OUT} = m\, c_V\, (T_4 - T_1)$	$V_4 = V_1$

Example 1.1

Calculate the states of the cold-air-standard Otto Cycle for a compression ratio of 12, with intake conditions at P = 1.0 bar and T = 300 K, assuming wide-open throttle. The heat release from combustion is Q/m = 1750 kJ/kg and the maximum volume in the cylinder is 1.0 L. What is the efficiency of the cycle?

SOLUTION: The initial state properties are given in the problem statement. To find the properties at the end of the compression stroke, the relationship for an isentropic compression of an ideal gas with constant specific heats can be used:

$$PV^k = C \tag{1.8}$$

Thus $P_1 V_1^k = P_2 V_2^k$, and the pressure at state 2 can be found from:

$$P_2 = P_1 \left(\frac{V_1}{V_2}\right)^k = P_1 (CR)^k = (1.0 bar)(12)^{1.4} = 32.4 bar$$

The temperature at state 2 can be found using the ideal gas law.

$$\frac{P_1 V_1}{T_1} = \frac{P_2 V_2}{T_2}$$

for a fixed mass of gas. Combining the ideal gas law with the isentropic compression relation, $P_1 V_1^k = P_2 V_2^k$, yields:

$$T_2 = T_1 (CR)^{k-1} = (300K)(12)^{0.4} = 811K$$

When using the ideal gas law, the temperature and pressure must always be expressed in absolute units. For constant volume combustion with constant specific heats, $Q_{chem} = m\ c_V\ \Delta T$.

$$\Delta T = \frac{Q}{m} \frac{1}{c_V} = \frac{1750 kJ/kg}{0.718 kJ/kg \cdot K} = 2437K$$

Thus the temperature at the end of combustion is $T_3 = T_2 + \Delta T = 811\ K + 2437\ K = 3248\ K$. Once again the ideal gas law can be used, this time to find the pressure. Since the combustion process occurs at constant volume, then

$$\frac{P_3}{T_3} = \frac{P_2}{T_2}$$

$$P_3 = (32.4bar)\frac{3248K}{811K} = 129.8bar$$

For the isentropic expansion of the power stroke the pressure-volume relationship is:

$$P_4 = P_3\left(\frac{V_3}{V_4}\right)^k = P_3\left(\frac{1}{CR}\right)^k = (129.8bar)\left(\frac{1}{12}\right)^{1.4} = 4.0bar$$

The final temperature is:

$$T_4 = T_3\left(\frac{1}{CR}\right)^{k-1} = (3248K)\left(\frac{1}{12}\right)^{0.4} = 1202K$$

The efficiency of the Otto cycle can be calculated from an energy balance using Equation 1.6.

$$\eta = \frac{Q_{IN} - Q_{OUT}}{Q_{IN}}$$

where $Q_{OUT} = c_V(T_4-T_1) = 0.718$ kJ/kg-K $(1202\text{ K} - 300\text{ K}) = 648$ kJ/kg.

$$\eta = \frac{1750 kJ/kg - 648 kJ/kg}{1750 kJ/kg} = 0.630 = 63.0\%$$

Table 1.2: Results for Example 1.1

State	Volume (L)	Pressure (bar)	Temperature (K)
1	1.0	1.0	300
2	0.083	32.4	811
3	0.083	129.8	3248
4	1.0	4.0	1202

Example 1.2

How does the efficiency of the Otto Cycle vary with the Compression Ratio?

SOLUTION: From Eq. 1.6, the efficiency is η = (Q_{IN} - Q_{OUT}) / Q_{IN}. If we assume constant specific heats, then $Q_{IN} = c_V (T_3 - T_2)$, and $Q_{OUT} = c_V (T_4 - T_1)$. Substituting into Eq. 1.6, the efficiency is:

$$\eta = \frac{T_3 - T_2 - (T_4 - T_1)}{T_3 - T_2} = 1 - \frac{T_4 - T_1}{T_3 - T_2}$$

From the isentropic compression and expansion relations, $T_3 = T_4(CR)^{k-1}$, and $T_2 = T_1(CR)^{k-1}$.

$$\eta = 1 - \frac{T_4 - T_1}{T_4 CR^{k-1} - T_1 CR^{k-1}} = 1 - \frac{T_4 - T_1}{(T_4 - T_1)CR^{k-1}}$$

Thus, the efficiency of the Otto Cycle with constant specific heats is:

$$\eta = 1 - \frac{1}{CR^{k-1}} \qquad (1.9)$$

The efficiency of the Otto cycle increases as the compression ratio is increased. In the limit as the compression ratio approaches infinity, and efficiency approaches 100%, as all of the heat release could be converted into work in an infinite expansion.

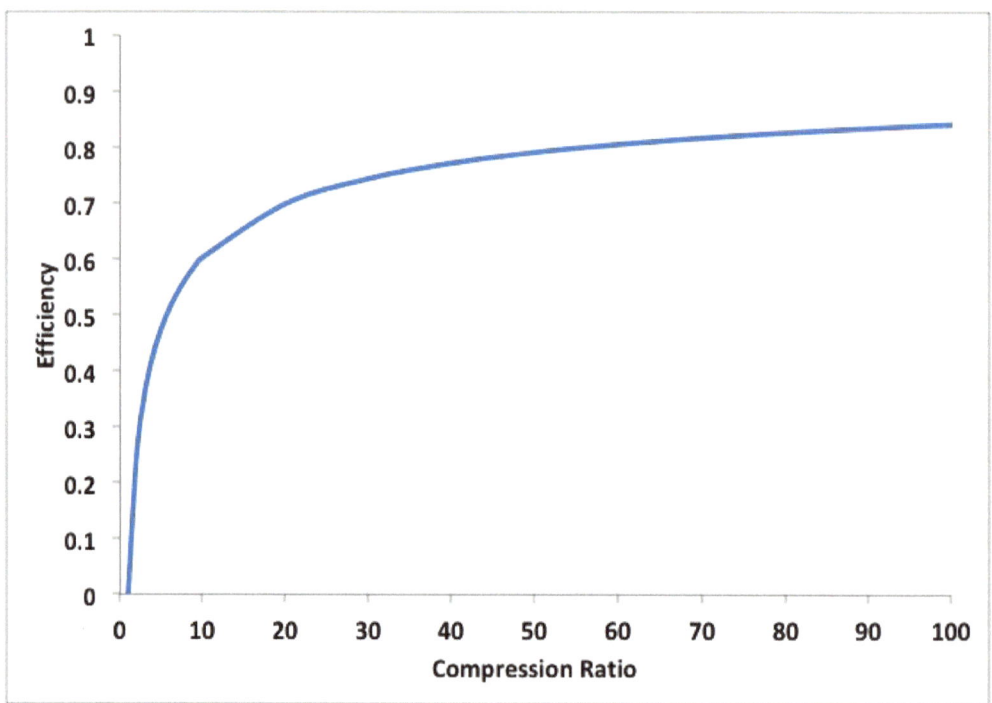

Figure 1.3: Efficiency of Otto Cycle as a function of Compression ratio.

1.3 Diesel Cycle

Rudolph Diesel (1858-1913) had studied thermodynamics at the Munich Technical University. There he learned about the work of Sadi Carnot (1796-1832), and the theoretical Carnot cycle, which is the cycle that gives the maximum possible efficiency for a heat engine operating between two thermal reservoirs at fixed temperatures. The Carnot Cycle is comprised of four processes:

1. Isothermal Compression
2. Isentropic Compression
3. Isothermal Expansion
4. Isentropic Expansion

The Carnot Cycle is shown on a P-V diagram in Figure 1.4.

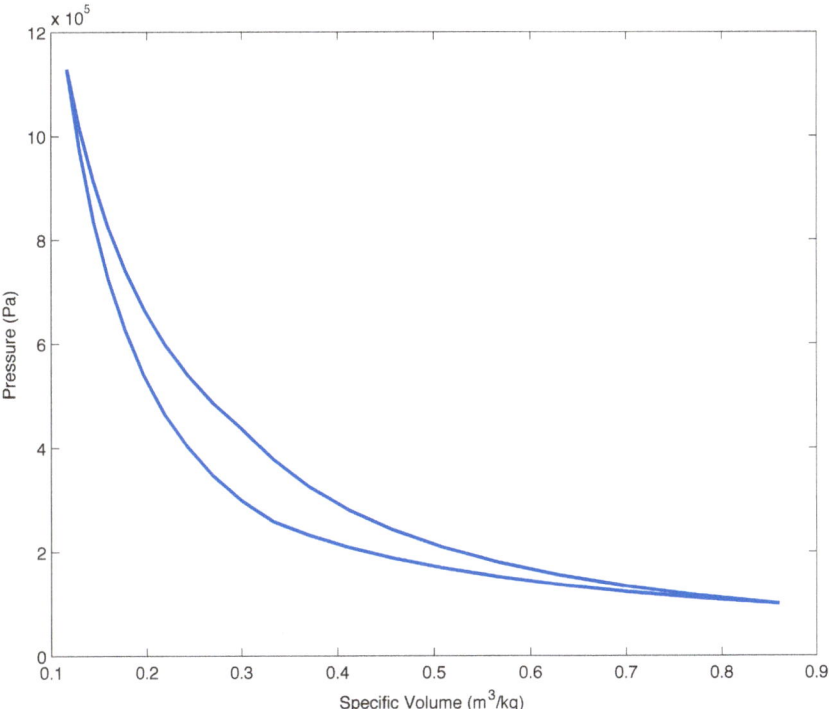

Figure 1.4: P-V diagram of Carnot cycle with air as the working fluid, with compression from initial condition of 1 bar, 300 K.

Diesel quickly realized that the isothermal compression process of Carnot's cycle was not practical for an engine operating at any significant speed, so he proposed that instead of separate isothermal and isentropic compression steps, to have a single isentropic compression from maximum to minimum volume of the piston-cylinder engine. He also wanted to avoid the knocking, or pre-ignition, problems of gasoline engines, so he wanted to keep the fuel out of the cylinder until the end of the compression stroke. In fact, he was relying on the high temperature of the compressed air to ignite the fuel without the need for a spark plug in his direction-injection, compression-ignition engine.

To achieve isothermal expansion, the rate of fuel injection would have to be slow, so that the temperature rise due to combustion exactly matched the temperature loss due to expansion. Diesel later discovered that while this strategy would lead to a high indicated thermal efficiency, it would not have an adequate mechanical efficiency, as the power lost due to friction would be nearly as great as the power generated from combustion. Diesel realized he would have to burn more fuel in each engine cycle to get an acceptable overall efficiency. He settled on an isobaric combustion process as a compromise. The isobaric (constant pressure) combustion process has the appealing feature that it would give high efficiency in a pressure-limited engine. The Diesel cycle is similar to the Otto cycle with only one change: combustion is modeled as a constant pressure process rather than a constant volume process. The cutoff ratio is the ratio of the volume at the end of the isobaric combustion process to the volume at the end of the compression stroke.

Table 1.3: Steps in the Diesel cycle.

Process	Thermodynamic relations	Volume relations
Isentropic Compression	$P_1 V_1^k = P_2 V_2^k$	$V_2 = V_1 /$ Compression ratio
Constant Pressure Heat Addition	$Q_{IN} = m\ c_P\ (T_3 - T_2)$	$P_3 = P_2$
Isentropic Expansion	$P_4 V_4^k = P_3 V_3^k$	$V_3 = V_2 *$ Cutoff ratio
Constant Volume Heat Rejection	$Q_{OUT} = m\ c_V\ (T_4 - T_1)$	$V_4 = V_1$

Example 1.3

Calculate the states of the cold-air-standard Diesel Cycle for a compression ratio of 12, with intake conditions at P = 1.0 bar and T = 300 K, maximum volume of 1.0 L, and heat release from combustion of Q/m = 1750 kJ/kg. What is the efficiency of the cycle?

SOLUTION: The isentropic compression stroke is the same as that of the Otto Cycle in Example 1.1. For constant volume combustion with constant specific heats, $Q_{chem} = m \, c_P \, \Delta T$.

$$\Delta T = \frac{Q}{m} \frac{1}{c_P} = \frac{1750 kJ/kg}{1.005 kJ/kg \cdot K} = 1741 K$$

Thus the temperature at the end of combustion is $T_3 = T_2 + \Delta T = 811 \, K + 1741 \, K = 2552 \, K$. Once again the ideal gas law can be used, this time to find the volume. Since the combustion process occurs at constant pressure, then

$$V_3 = V_2 \frac{T_3}{T_2} = (0.0833L) \frac{2552K}{811K} = 0.262L$$

The expansion ratio from state 3 to state 4 is then (1.0 L)/(0.262 L) = 3.81. The final pressure at the end of expansion is:

$$P_4 = P_3 \left(\frac{V_3}{V_4}\right)^k = (32.4 bar)(0.262)^{1.4} = 4.97 bar$$

and the final temperature is:

$$T_4 = T_3 \left(\frac{V_3}{V_4}\right)^{k-1} = (2552K)(0.262)^{0.4} = 1493K$$

The efficiency of the cycle can be calculated from an energy balance using Equation 1.6, where $Q_{OUT} = c_V(T_4-T_1) = 0.718$ kJ/kg-K (1493 K − 300 K) = 857 kJ/kg.

$$\eta = \frac{1750 kJ/kg - 857 kJ/kg}{1750 kJ/kg} = 0.510 = 51.0\%$$

Table 1.4 shows the calculated volume, pressure, and temperature for each state of the Diesel Cycle.

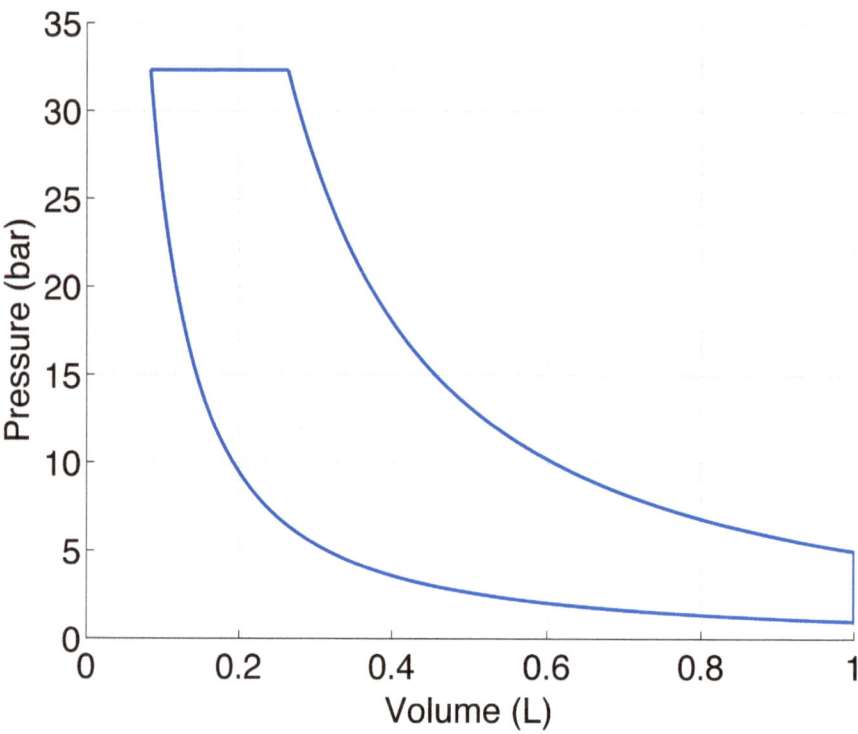

Figure 1.6: P-V diagram of diesel cycle with conditions corresponding to those in Example 1.3.

Table 1.4: Results for Example 1.3.

State	Volume (L)	Pressure (bar)	Temperature (K)
1	1.0	1.0	300
2	0.083	32.4	811
3	0.262	32.4	2552
4	1.0	5.0	1493

The efficiency of the Diesel cycle is 51%, compared to the 63% calculated for the Otto cycle in Example 1.1 for the same compression ratio and heat addition. So it can be seen that at the same compression ratio and for the same amount of heat released, the Otto Cycle is more efficient than the Diesel Cycle. Figure 1.7 shows a comparison of the Otto and Diesel cycles at the same compression ratio and for the same amount of heat release. It can be seen that the Otto Cycle has a larger area, and hence more work and a higher efficiency.

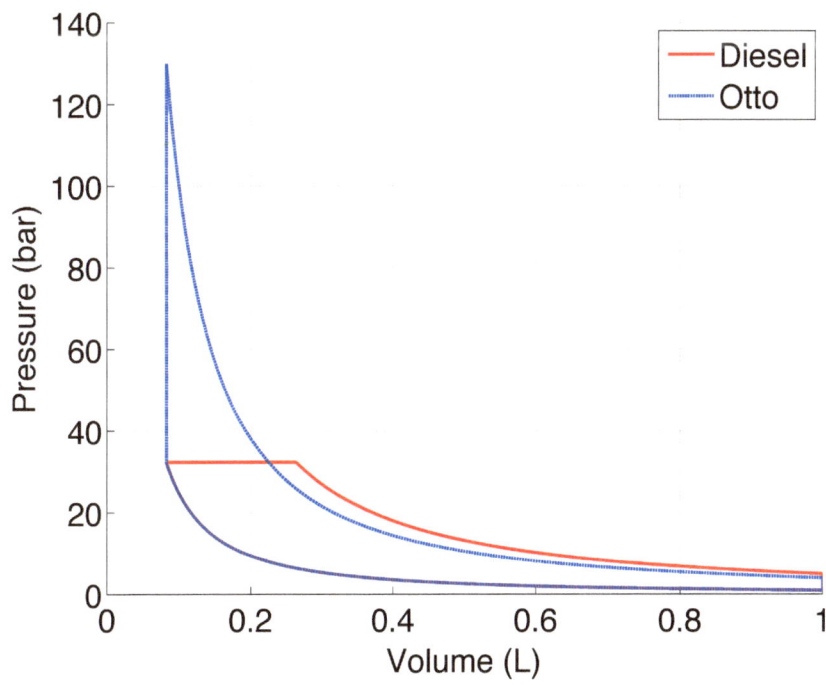

Figure 1.7: Comparison of P-V diagrams of Otto and Diesel cycles with same compression ratio and same amount of heat addition by combustion.

Example 1.4

Compare the efficiencies of the Otto and Diesel cycles for an engine that is limited to maximum pressure of 200 bar, with intake conditions at P = 1.0 bar and T = 300 K, maximum volume of 1.0 L, and heat release from combustion of Q/m = 1750 kJ/kg.

SOLUTION: For the Otto Cycle, the compression ratio will have to be adjusted such that the final pressure at the end of combustion will be 200 bar. With a fixed heat release due to combustion of 1750 kJ/kg, a compression ratio of 17.7 is necessary to give a maximum pressure of 200 bar. The pressure and temperature at the end of compression are 55.9 bar and 947 K. The pressure and temperature after combustion are 200 bar and 3384 K. After expansion, the pressure and temperature drop to 3.6 bar and 1072 K. The heat lost to the exhaust gases is $Q_{OUT} = c_V(T_4 - T_1)$ = 0.718 kJ/kg-K (1072 K − 300 K) = 554 kJ/kg, and the efficiency of the Otto Cycle is:

$$\eta = \frac{1750 kJ/kg - 554 kJ/kg}{1750 kJ/kg} = 0.683 = 68.3\%$$

For the Diesel Cycle, since combustion occurs at constant pressure, the initial compression stroke can be taken all the way to a pressure of 200 bar, so that the compression ratio is:

$$CR = \left(\frac{200 bar}{1 bar}\right)^{1/1.4} = 44.0$$

The temperature at the end of the compression stroke will be

$$T_2 = T_1 (CR)^{k-1} = (300K)(44)^{0.4} = 1363K$$

As in Example 1.3, the temperature rise due to combustion is 1741K, so that the temperature at the end of combustion is T_3 = 1363 K + 1741 K = 3104 K. The volume at the end of combustion can be calculated from:

$$V_3 = V_2 \frac{T_3}{T_2} = \left(\frac{1}{44}L\right)\frac{3104K}{1363K} = 0.052L$$

The expansion ratio from state 3 to state 4 is then (1.0 L)/(0.052 L) = 19.3. The final pressure and temperature at the end of expansion are:

$$P_4 = P_3 \left(\frac{V_3}{V_4}\right)^k = (200 bar)(0.052)^{1.4} = 3.2 bar$$

$$T_4 = T_3 \left(\frac{V_4}{V_3}\right)^{k-1} = (3104K)(0.052)^{0.4} = 950K$$

Q_{OUT} = $c_V(T_4-T_1)$ = 0.718 kJ/kg-K (950 K – 300 K) = 467 kJ/kg, and the efficiency of the Diesel Cycle is:

$$\eta = \frac{1750 kJ/kg - 467 kJ/kg}{1750 kJ/kg} = 0.733 = 73.3\%$$

Figure 1.8 shows a comparison of the Otto and Diesel cycles for the same maximum cylinder pressure and for the same amount of heat release. It can be seen that the Diesel Cycle has a larger area, and hence more work and a higher efficiency. For a pressure-limited engine, the Diesel cycle is more efficient than the Otto Cycle, for the same amount of energy input.

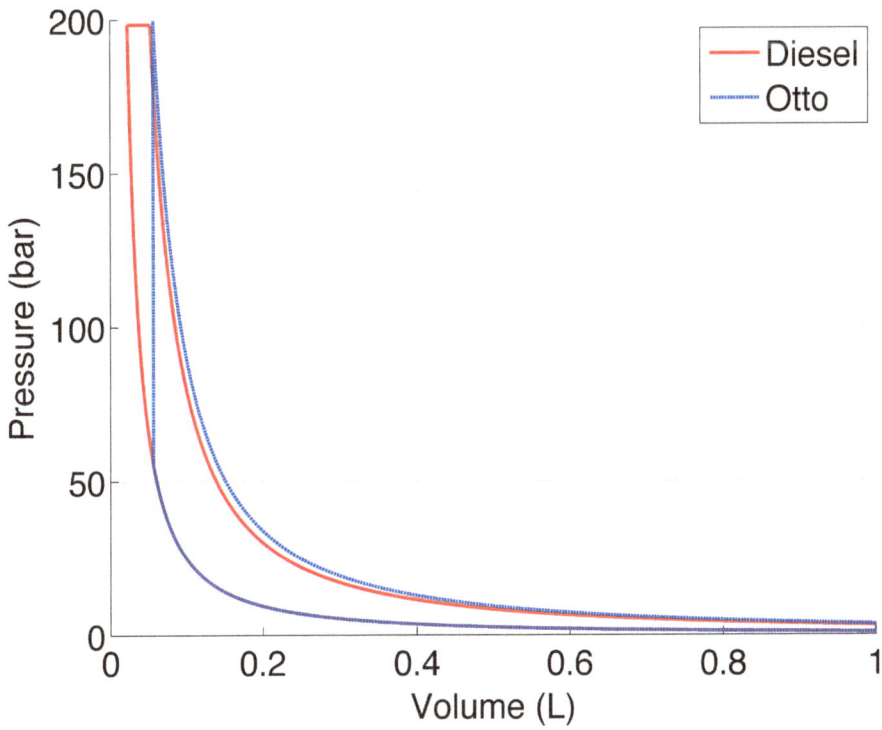

Figure 1.8: Comparison of P-V diagrams of Otto and Diesel cycles with same maximum pressure and same amount of heat addition by combustion.

Real diesel engines are usually more limited by the maximum allowable working pressure than they are by the compression ratio. The compression ratio itself is limited by the need to have some clearance volume so the valves and fuel injector tip will not hit the piston. Gasoline engines are compression-ratio limited because of *knocking*, which is premature detonation of the fuel-air mixture before the flame front initiated by the spark plug reaches it. Anti-knock additives are added to gasoline to allow for higher compression ratios. The early gasoline engines of the late 1800s and early 1900s were limited to compression ratios of about 4:1 to avoid knock. As a result, they had correspondingly low efficiencies. Tetra-ethyl lead was introduced in the 1920s, and then was phased out by 1996 over concerns about lead emissions. To replace lead, other additives, such as butane, aromatics, alcohols, and ethers have been used. Iodine and aniline were used before lead as antiknock additives to gasoline. The octane rating on the gasolines sold at fuel stations is supposed to be a measure of a fuel's resistance to knocking, with higher octane numbers representing better knocking resistance and the potential to use higher compression ratios. However, a recent paper suggests that the octane ratings at the pump are not a good measure of knocking resistance, since the testing procedures for measuring the octane number were developed 80 years ago and are not of relevance to modern engines [Mittal09]. Typical compression ratios for diesel engines are in the range of 16:1 to 23:1. This would correspond to theoretical efficiencies of 67.0% to 71.5% for the Otto cycle, or efficiencies of 57.3% to 64% using the Diesel cycle with heat input of 1750 kJ/kg, as in Example 1.4. The efficiency of real engines will be less than that of the theoretical cycles at the same compression ratio due to heat transfer losses and other irreversibilities, and Figure 1.9 shows the efficiency for production on highway diesel truck engines.

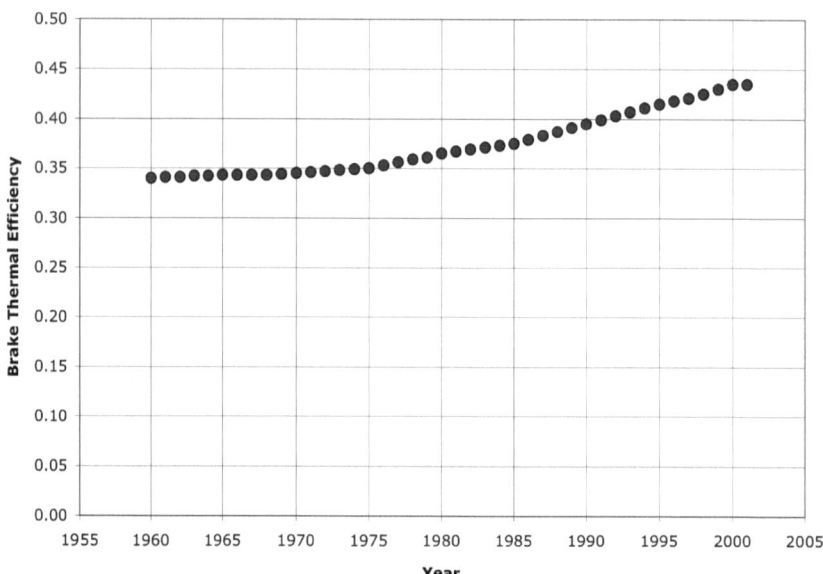

Figure 1.9: First law efficiency for on highway heavy-duty diesel truck engines. Data from [Aneja09].

1.4 Dual Cycle

The dual cycle is an attempt to more closely model experimentally seen P-V traces from internal combustion engines with a simple thermodynamic cycle, and it includes 5 steps instead of 4. One problematic feature of the theoretical Dual cycle is the necessity of determining how to split the heat input between the constant volume process and the constant pressure process.

Table 1.5: Steps in the Dual Cycle.

Process	Thermodynamic relations	
Isentropic Compression	$P_1V_1^k = P_2V_2^k$	$V_2 = V_1$ / Compression ratio
Constant Volume Heat Addition	$Q_{IN} = m\ c_V\ (T_3 - T_2)$	$V_3 = V_2$
Constant Pressure Heat Addition	$Q_{IN} = m\ c_P\ (T_4 - T_3)$	$P_4 = P_3$
Isentropic Expansion	$P_5V_5^k = P_4V_4^k$	$V_4 = V_3$ * Cutoff ratio
Constant Volume Heat Rejection	$Q_{OUT} = m\ c_V\ (T_5 - T_1)$	$V_5 = V_1$

None of the theoretical cycles (Otto, Diesel, or Dual) do a good job of matching the actual P-V relationship in real production engines. In real engines, while the compression process is fairly close to isentropic, the actual combustion process is neither isochoric nor isobaric, and there are significant heat losses during the expansion stroke. In order to more closely model real engine performance, an engine cycle simulation using a computer code must be used, which is discussed in the next section.

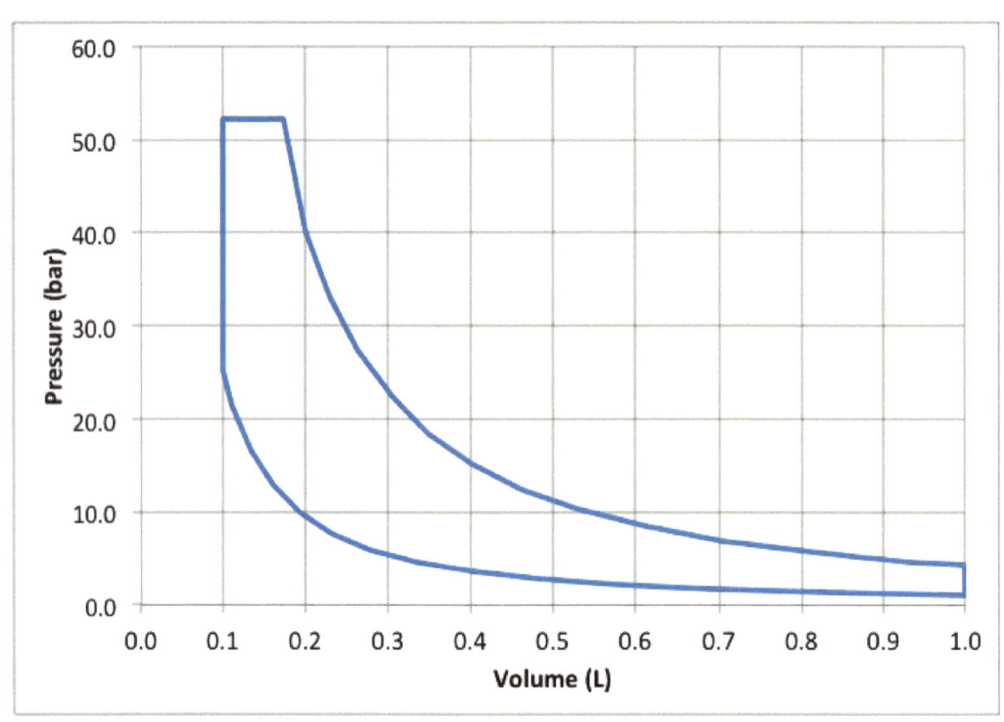

Figure 1.10: Diagram of Dual Cycle for compression ratio of 10.

1.5 Engine Cycle Simulations

In order to perform a cycle analysis that includes heat loss to the walls and accounts for the finite time duration of combustion, a computer simulation must be used.

ASSUMPTIONS

1. Constant specific heats
2. Zero-dimensional flow conditions inside the cylinder
3. The gas in the cylinder moves through equilibrium states
4. No gas leakage through valves or piston rings, so that the mass is constant
5. Uniform crank speed
6. The cylinder gas is air and it obeys the ideal gas law

ANALYSIS

Starting with the general form of the first law for a closed, transient system:

$$\frac{dE_{CV}}{dt} = \dot{Q} - \dot{W} = \frac{dQ}{dt} - \frac{dW}{dt} \tag{1.10}$$

Under the assumptions stated previously of constant specific heats we can write:

$$E = m\,u = m\,c_V\,T \tag{1.11}$$

Boundary work can be calculated from:

$$W = \int P\,dV$$

taking the derivative of both sides of the equation, the differential amount of work, dW, can be seen to be:

$$dW = P\,dV \tag{1.12}$$

Substituting for E and W into the first law, we obtain:

$$\frac{d(m c_V T)}{dt} = \frac{dQ}{dt} - \frac{(P\,dV)}{dt}$$

With the assumption of constant specific heats, this expression can be written as:

$$c_V \frac{d(mT)}{dt} = \frac{dQ}{dt} - P\frac{dV}{dt}$$

For the time being we will leave the heat transfer term, dQ/dt, as it is. With further knowledge of heat transfer we could substitute an appropriate correlation for heat loss to the walls. The ideal gas law is PV = mRT, so we can substitute with mT = PV/R to eliminate temperature from the equation. (Note that here for simplicity R represents the gas constant for air on a per mass basis.)

$$\frac{c_V}{R}\frac{d(PV)}{dt} = \frac{dQ}{dt} - P\frac{dV}{dt}$$

Using the product rule for the derivative of a product:

$$\frac{c_V}{R}\left[V\frac{dP}{dt} + P\frac{dV}{dt}\right] = \frac{dQ}{dt} - P\frac{dV}{dt}$$

Since dV/dt will be known from engine speed and geometry, solve for dP/dt.

$$\frac{dP}{dt} = \frac{1}{V}\frac{R}{c_V}\frac{dQ}{dt} - \frac{P}{V}\left(1 + \frac{R}{c_V}\right)\frac{dV}{dt}$$

Since $R = c_P - c_V$ and $k = c_P / c_V$, then $R/c_V = k - 1$. The instantaneous crank angle, θ, is related to the time, t, through

$$\theta = \omega t \tag{1.13}$$

Assuming the engine speed, ω, is constant, we can also divide through by ω to obtain:

$$\frac{dP}{d\theta} = \frac{(k-1)}{V}\frac{dQ}{d\theta} - k\frac{P}{V}\frac{dV}{d\theta} \tag{1.14}$$

ENGINE MOTION

From the geometry of a reciprocating automobile engine with a rotating crankshaft and a connecting rod linking piston and crank:

$$V(\theta) = \frac{V_{DISP}}{CR-1} + \frac{V_{DISP}}{2}\left[R + 1 - \cos(\theta) - \sqrt{R^2 - \sin^2(\theta)}\right] \tag{1.15}$$

where R is the ratio of connecting rod length to crank radius. Crank radius is half the stroke length. The equation for the derivative of volume is given by:

$$\frac{dV}{d\theta} = \frac{V_{disp}}{2}\sin\theta\left(1 + \frac{\cos\theta}{\sqrt{R^2 - \sin^2\theta}}\right) \quad (1.16)$$

NUMERICS

To solve this differential equation we will employ the first order explicit *Euler method*:

$$P_{n+1} = P_n + \frac{dP}{d\theta} \cdot \Delta\theta \quad (1.17)$$

where

$$\frac{dP}{d\theta} = f(P_n, V(\theta), V'(\theta), Q'(\theta))$$

For $n = 1$ (initial condition), 2, 3, 4, ..., n_{max}.

ALGORITHM 1.1 – ENGINE CYCLE SIMULATION

1. Define Engine Operating Parameters (CR, V_{DISP}, N, Q_{IN})
2. Set numerical time step, $\Delta\theta$ (should be no greater than 1.0 CAD)
3. Start the simulation at -180 CAD (BDC)
4. Apply Equation 1.17 in a loop until the end of the expansion stroke at +180 CAD

The engine cycle simulation can be implemented in MATLAB, Python or other programming environment.

Example 1.5:

Use the Engine Cycle Simulation Algorithm to Re-create the Otto Cycle for the same conditions as Example 1.1 (CR = 12, constant specific heats, no heat transfer losses, instantaneous combustion of Q/m = 1750 kJ/kg). Also run a simulation for a motoring case (no combustion).

SOLUTION: The MATLAB code used is included in an Appendix at the end of this book, and results are plotted in Figure 1.11. A time step 0.1° was used. For the motoring case, the computed pressure at TDC is 32.2 bar, which is within 0.6% of that computed in Example 1.1 using the isentropic relations (32.4 bar).

Example 1.6:

Repeat the combustion of Example 1.5, but with a finite time duration of combustion. Assume heat release due to combustion takes place at constant rate starting at TDC, spread out over a time duration of 40 crank angle degrees.

SOLUTION: MATLAB generated plot shown in Fig. 1.11. The efficiency of the cycle is computed to be 54.7%, which is lower than the 63% computed for the Otto cycle. When the combustion heat is released instantaneously at TDC, it has the full 180° of the expansion stroke to do useful work, while for the finite duration heat release spread out over 40 degrees, there is only about 160° available on average for work to be done, and the efficiency is reduced by approximately that ratio. If we were to spread out the combustion heat release over 80°, from 0 to 80 degrees after top dead center, the efficiency drops to 41.7%

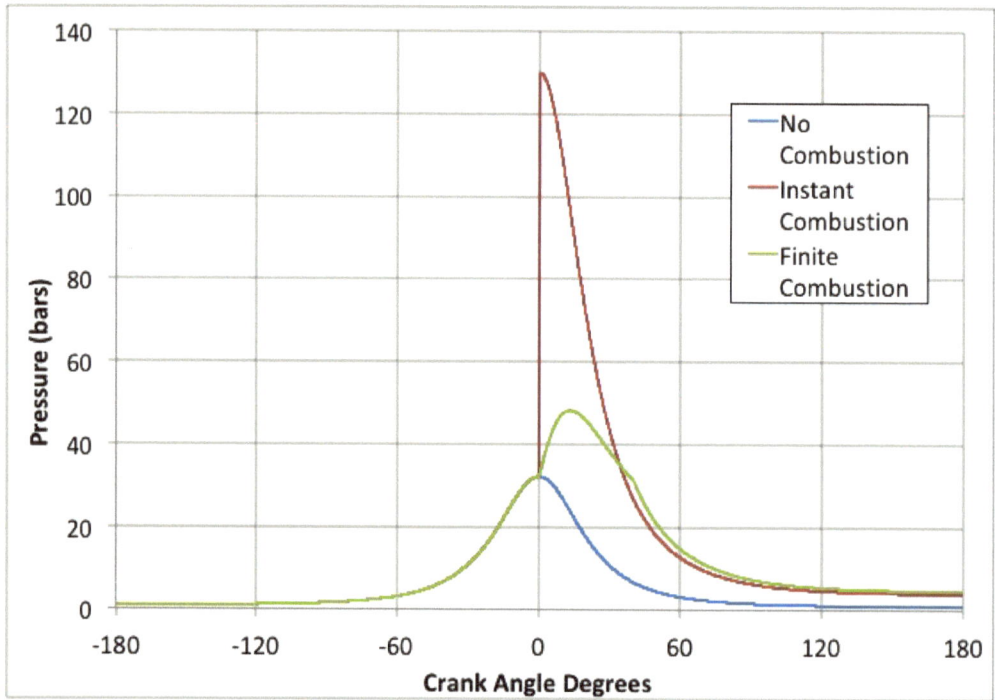

Figure 1.11: Pressure vs. Crank angle curve for conditions of Examples 1.5 & 1.6, compared to pressure curve for instantaneous combustion.

Figure 1.12 shows how the cycle efficiency varies with the timing of the start of combustion, for a combustion duration of 80 crank angle degrees. Real engines use a Spark Advance to obtain the maximum efficiency, in which spark ignition starts before top-dead-center (TDC) at 0°. Even though there is some negative work done as the piston pushes upwards against the combusting gases, the net work increases, since there is more time on the downstroke for the hot gases to do work when combustion starts earlier.

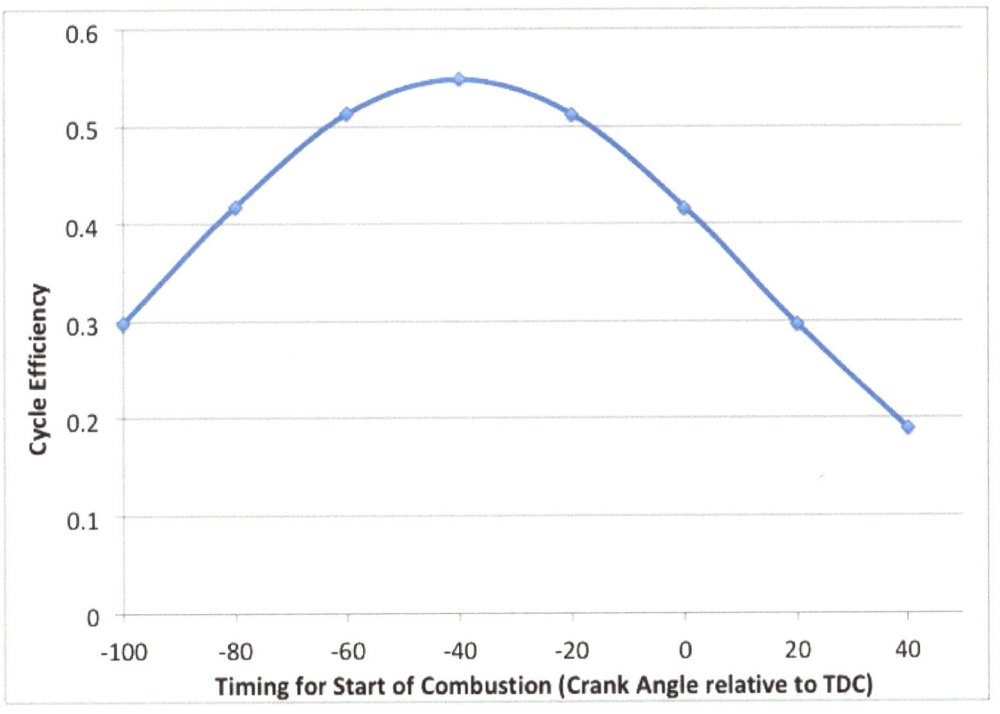

Figure 1.12: Efficiency as a function of start of combustion timing.

Heat transfer in engines

For heat transfer losses to the cylinder walls, convective heat transfer dominates:

$$Q = h A (T - T_{wall}) \tag{1.18}$$

Where h is the convective heat transfer coefficient. Heat transfer correlations are generally given in the form of Nu = f(Re), where the Nusselt number is defined as:

$$Nu = \frac{hL}{k} \tag{1.19}$$

where k is the thermal conductivity of the cylinder gases, and the Reynolds number is defined as:

$$Re = \frac{\rho V L}{\mu} = \frac{VL}{\nu} \tag{1.20}$$

Where ν is the kinematic viscosity of the air. Typically the cylinder internal bore diameter, B, is used for the length scale, L. The velocity normally scales with the mean piston speed, which is given by:

$$V_p = 2 N s \tag{1.21}$$

Where N is the engine rotational speed, normally given in units of RPM, and s is the stroke length. N will need to be converted from per RPM to revolutions per second. For instantaneous heat transfer, the Annand correlation for 4-stroke engines, neglecting radiation heat transfer, is [Annamd63]:

$$Nu = 0.49\, Re^{0.7} \tag{1.22}$$

At STP, the properties of air are: k = 0.026 W/m-K, and ν = 1.5 × 10^{-5} m^2/s. However, both of these values increase with increasing temperature. The thermal conductivity scales with $T^{0.75}$ and the absolute viscosity scales with $T^{0.62}$ [Heywood88]. Another correlation for engine heat transfer is given by [Eichelberg39]:

$$h = 2.43 V_p^{1/3} \sqrt{PT} \tag{1.23}$$

while the Eichelberg correlation is not dimensionally homogenous, it is easy to implement in a computer code due to its simplicity. If the mean piston speed is in units of m/s, the instantaneous pressure in bars, and the instantaneous temperature in K, then the convective heat transfer coefficient given in Eq. 1.19 will have units of $W/m^2\text{-}K$.

Example 1.7

For an engine at N = 1800 RPM and stroke of s = 10 cm = 0.1 m, a bore of B = 10 cm = 0.1 m, intake conditions at 1 bar and 300 K, and a compression ratio of 10, estimate the heat transfer coefficient at TDC, assuming combustion starts shortly after TDC.

SOLUTION: The mean piston speed is V_P = 2*1800/60*0.1 = 6 m/s. Assuming isentropic compression to TDC, the TDC temperature will be $300*(10)^{0.4}$ = 300*2.51 = 753 K. At this temperature we can estimate the thermal conductivity and viscosity as:

$$k = 0.026 \frac{W}{m \cdot K} \left(\frac{753K}{295K}\right)^{0.75} = 0.053 \frac{W}{m \cdot K}$$

$$\mu = 1.8 \times 10^{-5} \frac{kg}{m \cdot s} \left(\frac{753K}{295K}\right)^{0.62} = 3.2 \times 10^{-5} \frac{kg}{m \cdot s}$$

If we take an average viscosity value of 10^{-4} m²/s, the Reynolds number is: Re = (12 kg/m³)(6 m/s)(0.1 m)/(0.000032 kg/m-s) = 225,000. Using the Annand correlation the Nusselt number is:

$$Nu = 0.49(225,000)^{0.7} = 2730$$

the convective heat transfer coefficient is:

$$h = Nu\frac{k}{B} = 2730 \frac{0.053 W/m \cdot K}{0.1m} = 1450 \frac{W}{m^2 K}$$

If we use the Eichelberg correlation, we also need to know the pressure at TDC, which is P = 1.0 bar*$(10)^{1.4}$ = 25.1 bar.

$$h = 2.43\left(6\frac{m}{s}\right)^{1/3} \sqrt{(25.1bar)(753K)} = 607 \frac{W}{m^2 K}$$

The Annand correlation generally predicts higher heat transfer coefficients than the Eichelberg correlation.

Example 1.8

Use an engine cycle simulation to model an engine with a compression ratio of 12, with intake conditions at P = 1.0 bar and T = 300 K, maximum volume of 1.0 L, and heat release from combustion of Q/m = 1750 kJ/kg, including heat transfer losses and finite combustion. What is the efficiency of the cycle?

SOLUTION: The MATLAB code used to perform the simulation is included at the end of this book. Figure 1.13 shows a pressure vs. volume plot and Figure 1.14 shows a pressure vs. crank angle plot. With the heat losses included for a 40 duration of combustion the cycle efficiency is 45.8% (compared to 54.7% with no heat losses). With an 80 duration of combustion and heat losses included the cycle efficiency is 34.5%, which is a realistic number for production engines.

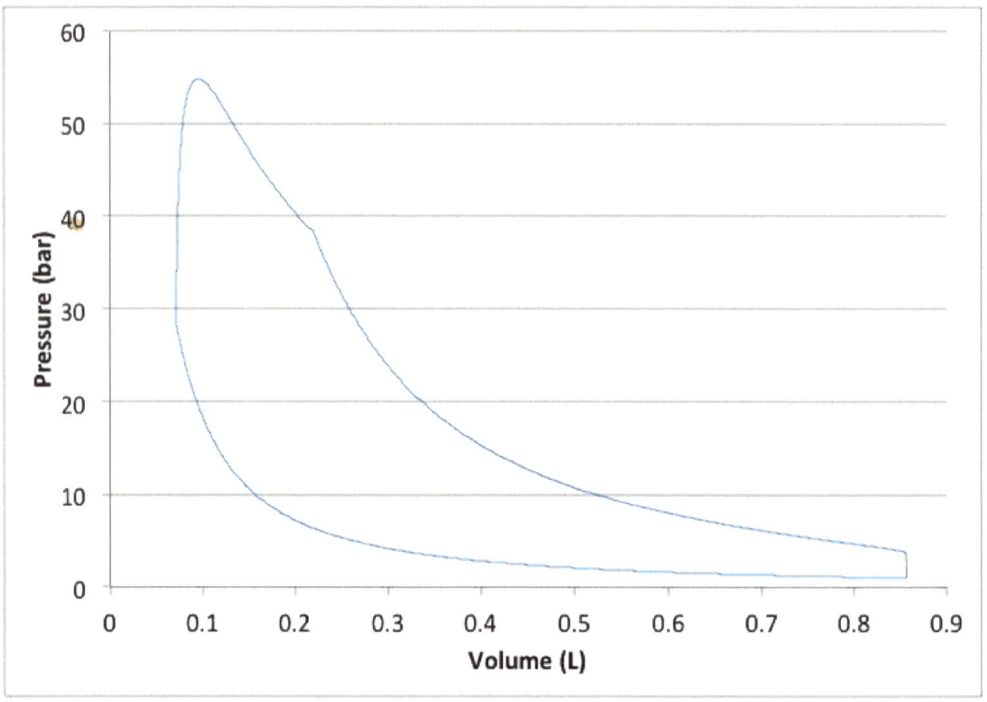

Figure 1.13: P-V diagram from engine cycle simulation with finite duration of combustion and heat losses to the walls.

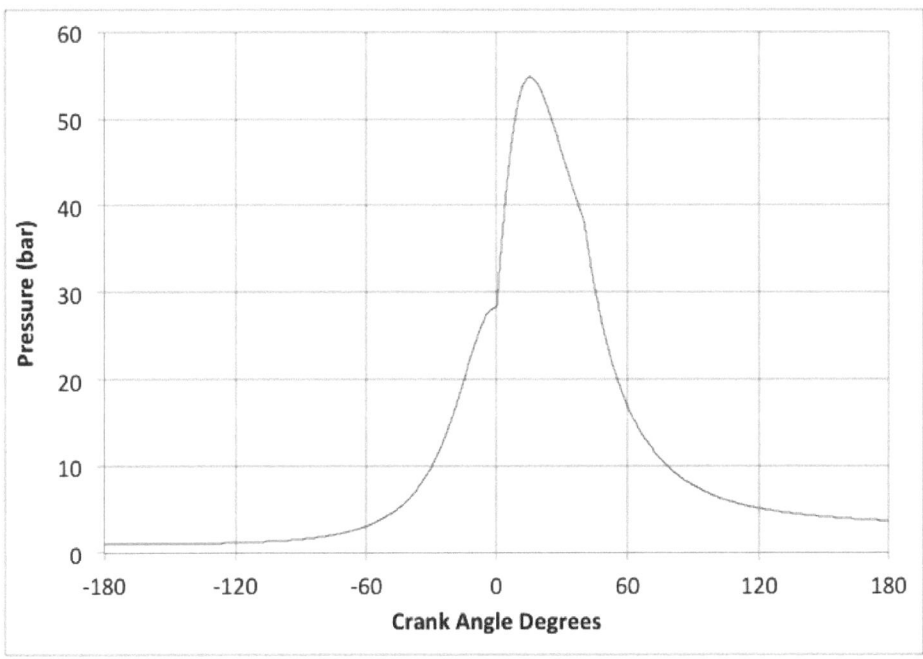

Figure 1.14: Pressure vs. crank angle graph from engine cycle simulation with finite duration of combustion and heat losses to the walls.

Variable Properties

With cold-air standard cycles (Otto, Diesel, Dual), it is necessary to assume the properties of the working gas are constant if the calculations are to be performed by hand. Thus properties of air at standard conditions are often assumed. In the real engine the gas composition changes as well as the operating temperature, both of which affect the properties of the working fluid (heat capacity, thermal conductivity). With cycle simulations performed on a computer we can relax these assumptions. The chemical composition of the post-combustion gases is discussed in more detail in Section 1.9. For complete stoichiometric combustion, the products of combustion are primarily nitrogen, carbon dioxide, and water vapor. This mixture has properties that are not all that different from that of air (nitrogen and oxygen mixture). The main different being the heat capacity is slightly higher for exhaust gases (about 5% higher). If this level of error is acceptable, it is often convenient to use the properties of air for that of the post-combustion gases in some cycle simulations for simplicity. The affects of temperature on the specific heat ratio, k, can be modeled with a linear fit for air from 150 K to 3000 K as:

$$k = 1.4224 - 0.00008T$$

for the temperature, T, in units of Kelvin.

Intake and Exhaust flow

The theoretical Otto and Diesel cycles, as well as the engine cycle simulation code, only model two strokes of a four-stroke engine. The compression and expansion (power) strokes are included, while the intake and exhaust strokes are omitted. The engine cycle simulation can be modified however to include all four strokes. Starting with the ideal gas law for the gases in the cylinder:

$$P = \frac{mRT}{MV} \tag{1.24}$$

Taking the derivative of both sides with respect to crank angle, θ, and assuming isothermal flow gives:

$$\frac{dP}{d\theta} = \frac{RT}{M}\frac{d}{d\theta}\left(\frac{m}{V}\right) = \frac{RT}{MV}\left(\frac{dm}{d\theta} - \frac{m}{V}\frac{dV}{d\theta}\right) \tag{1.25}$$

The assumption of isothermal flow will be valid for low engine speeds when the flow through the valves is at low Mach number. The mass flow rate into the cylinder from the intake manifold is:

$$\dot{m} = \frac{dm}{dt} = \omega\frac{dm}{d\theta} = \rho A V = \rho A C_d \sqrt{\frac{2(P_{int} - P(\theta))}{\rho}} \tag{1.26}$$

where A is the available flow area of the intake valves, C_d is the discharge coefficient for the valves, and P_{int} is the pressure in the intake manifold. For flow out of the cylinder during the exhaust stroke,

$$\dot{m} = \frac{dm}{dt} = \rho A V = -\rho A C_d \sqrt{\frac{2(P(\theta) - P_{exh})}{\rho}} \qquad (1.27)$$

For Ma > 0.3, the assumption of isothermal incompressible flow is not valid, but isentropic flow can be assumed instead. For this case the mass flow rate into the engine cylinder during intake can be written as:

$$\dot{m} = \rho_o A C_d \sqrt{k(R/M)T} \left(\frac{P_{int}}{P(\theta)}\right)^{\frac{1}{k}} \sqrt{2 \frac{1}{k-1}\left[\left(\frac{P_{int}}{P(\theta)}\right)^{\frac{k-1}{k}} - 1\right]} \qquad (1.28)$$

where T is the temperature in the intake manifold, ρ_o is the stagnation density of air flowing through the intake, and k is the specific heat ratio of the gas, equal to 1.4 for air at standard conditions. Note that the speed of sound in an ideal gas is given by:

$$a = \sqrt{k(R/M)T} \qquad (1.29)$$

For air at standard conditions the speed of sound is:

$$a = \sqrt{1.4 \left(\frac{8314 J/kmol \cdot K}{29 kg/kmol}\right)(298K)} = 345 m/s$$

Also note that for compressible flow, if the pressure ratio exceeds a critical value of:

$$\frac{P_{int}}{P(\theta)} > \left(\frac{k+1}{2}\right)^{\frac{k}{k-1}} = \left(\frac{1.4+1}{2}\right)^{\frac{1.4}{1.4-1}} = 1.89$$

Then the flow will choke, achieving sonic velocity (Ma = 1) at the valve. Once the flow is choked, the mass flow rate will not increase even if the manifold pressure is increased. For choked flow through a valve, the mass flow rate is given by:

$$\dot{m} = \rho_o A C_d \sqrt{kRT} \left(\frac{2}{k+1}\right)^{\frac{k+1}{2(k-1)}} \qquad (1.30)$$

Example 1.9

Compute the intake stroke for the conditions of Example 1.8, using an intake valve area of 0.000125 m^2 and discharge coefficient of C_d = 0.75, with intake manifold pressure of 1.0 bar (corresponding to wide-open-throttle condition).

SOLUTION: The MATLAB code used is included at the end of this book, and the instantaneous pressure in the cylinder as a function of cylinder volume during the intake stroke is shown in Figure 1.15.

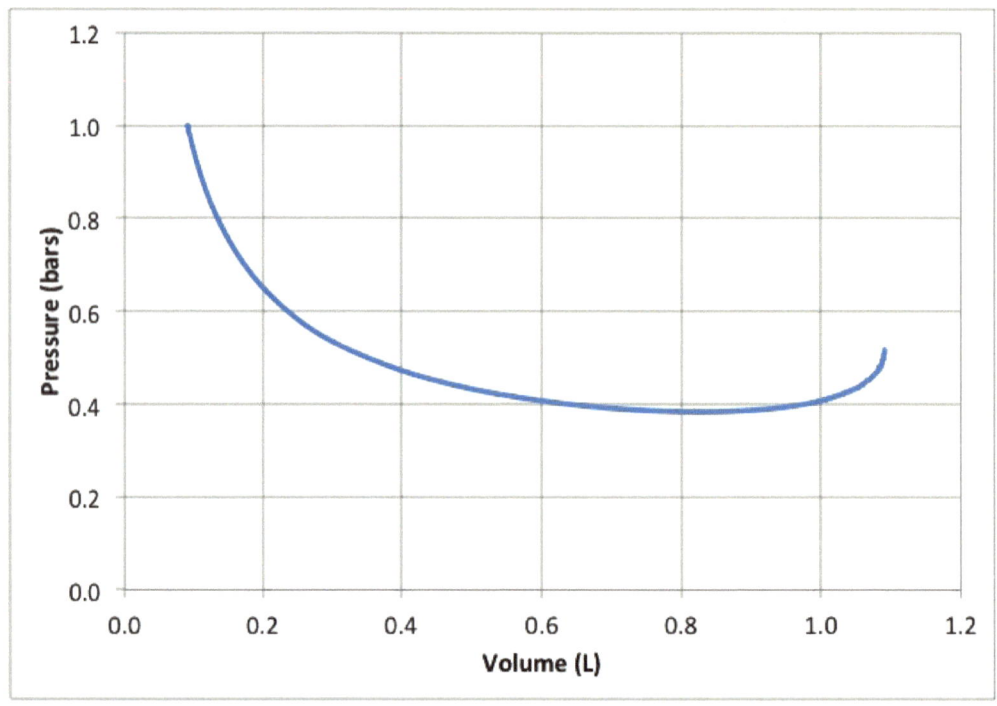

Figure 1.15: P-V diagram from engine cycle simulation of intake stroke of a 4-stroke engine.

The volumetric efficiency is defined as the ratio of air that is actually inducted to the engine to the amount that could be obtained if the displacement volume of the engine were completely filled with air at the density of air in the intake manifold.

$$\eta_{vol} = \frac{\dot{m}_{air}}{\rho_i V_{Disp} N/2} \tag{1.31}$$

where N is the engine speed, in revolutions per second, and the factor of 2 is to account for that in 4-stroke engines there is one intake stroke per cycle, or per 2 revolutions of the crankcase. The volumetric efficiency generally decreases with increasing engine speed.

1.6 Brayton Cycle and Gas Turbines

George Brayton (1839-1892) originally developed his cycle to model the reciprocating piston engines he was working on. The Brayton Cycle has come to be used to model open flow gas turbine engines, with continuous combustion.

Table 1.6: Steps in Brayton Cycle:

Process	Thermodynamic relations	Pressure relations
Isentropic Compression	$T_2 = T_1 * (P_2/P_1)^{(k-1)/k}$	$P_2 = P_1 *$ Compressor P ratio
Constant Pressure Heat Addition	$Q_{IN} = m\, c_P\, (T_3 - T_2)$	$P_3 = P_2$
Isentropic Expansion	$T_4 = T_3 * (P_4/P_3)^{(k-1)/k}$	$P_4 = P_3 /$ Compressor P ratio
Constant Pressure Heat Rejection	$Q_{OUT} = m\, c_P\, (T_4 - T_1)$	$P_4 = P_1$

It can be shown the efficiency of the simple Brayton cycle increases as the pressure ratio increases. An important parameter in analyzing gas turbines is the back work ratio, which is the percentage of work generated by the turbine that goes to power the compressor. Gas turbines can be used either for aircraft propulsion or stationary power generation. In propulsion applications all of the turbine power must go to the compressor or other thrust generating devices, such as fans or propellers. More complex Brayton cycles for power plants can be constructed when the use of regenerators or multiple turbines is to be modeled. Coal-burning and nuclear power plants are normally modeled with a Rankine cycle, which involves phase change of the working fluid and hence is not included in this book.

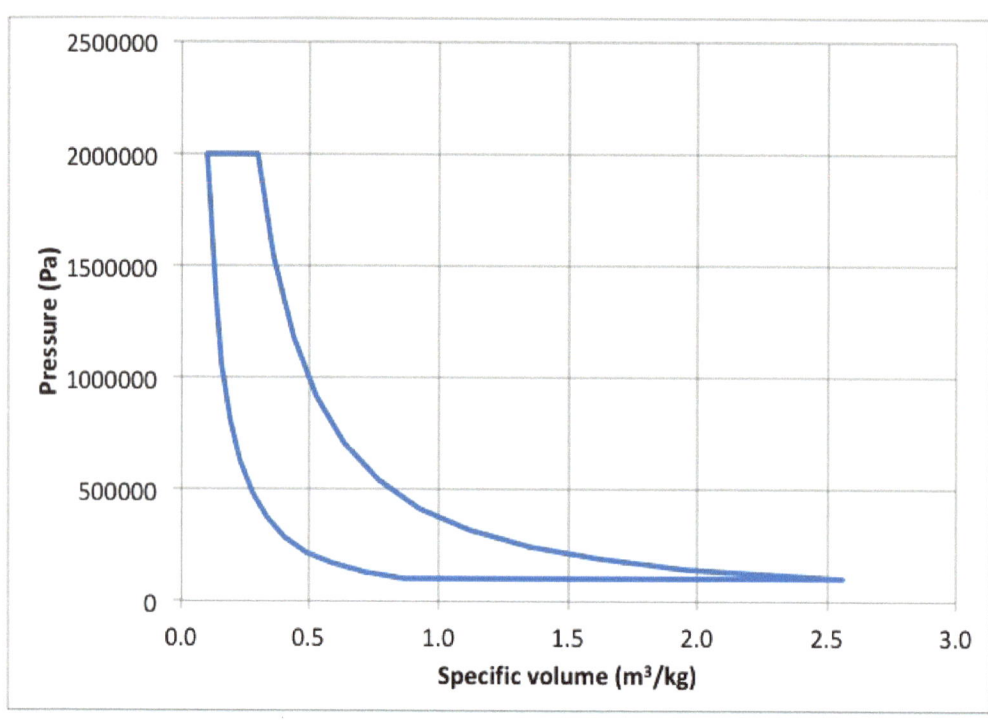

Figure 1.16: P-V diagram of Brayton cycle with a compressor pressure ratio of 20 and heat addition of 1400 kJ/kg.

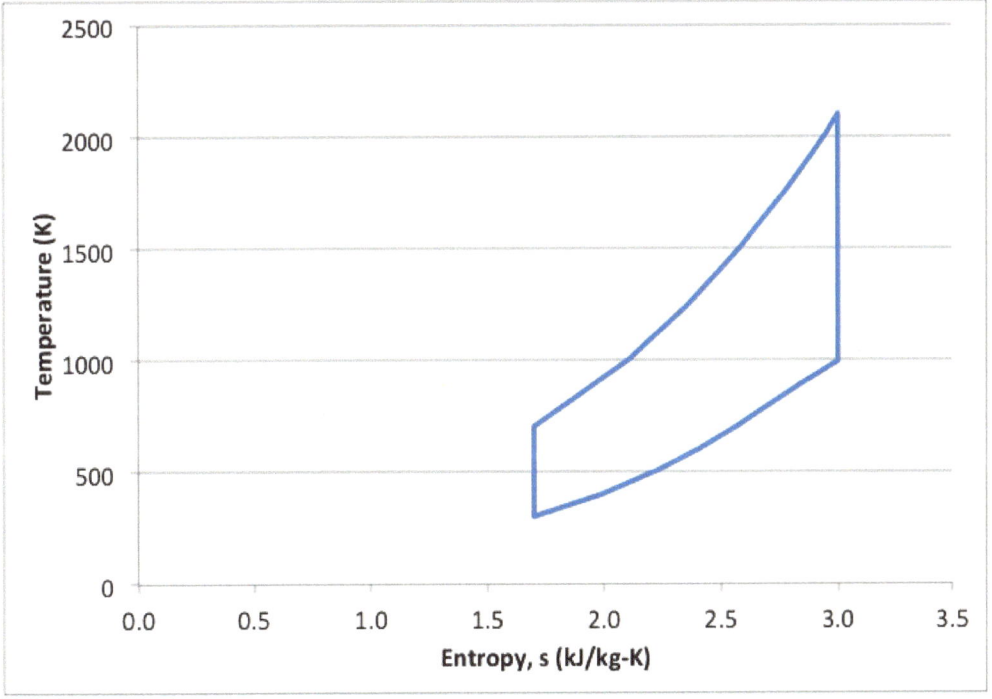

Figure 1.17: T-s diagram of Brayton cycle

The efficiency of Brayton cycle can be defined in the same way as the efficiency for any heat engine:

$$\eta = \frac{W_{net}}{Q_{IN}} = \frac{Q_{IN} - Q_{OUT}}{Q_{IN}}$$

If we assume constant specific heats, then $Q_{IN} = m\, c_P\, (T_3 - T_2)$, and $Q_{OUT} = m\, c_P\, (T_4 - T_1)$. Substituting these relations in to the formula for efficiency, we have:

$$\eta = \frac{T_3 - T_2 - (T_4 - T_1)}{T_3 - T_2} = 1 - \frac{T_4 - T_1}{T_3 - T_2}$$

Defining the compressor pressure ratio as PR = P_2/P_1, we have $T_2 = T_1*(PR)^{(k-1)/k}$ and $T_3 = T_4*(PR)^{(k-1)/k}$. Substituting into the previous expression for efficiency,

$$\eta = 1 - \frac{T_4 - T_1}{T_4 PR^{(k-1)/k} - T_1 PR^{(k-1)/k}} = 1 - \frac{T_4 - T_1}{(T_4 - T_1) PR^{(k-1)/k}}$$

The first law efficiency for the Brayton cycle with constant specific heats is:

$$\eta = 1 - \frac{1}{PR^{\frac{k-1}{k}}} \qquad (1.32)$$

Therefore it can be seen that the efficiency of the ideal cold-air standard Brayton cycle increases with an increase in the compressor pressure ratio. Figure 1.18 shows the efficiency of the ideal Brayton cycle as a function of pressure ratio, and Figure 1.19 shows the historical trends in pressure ratio increases in production jet aircraft engines.

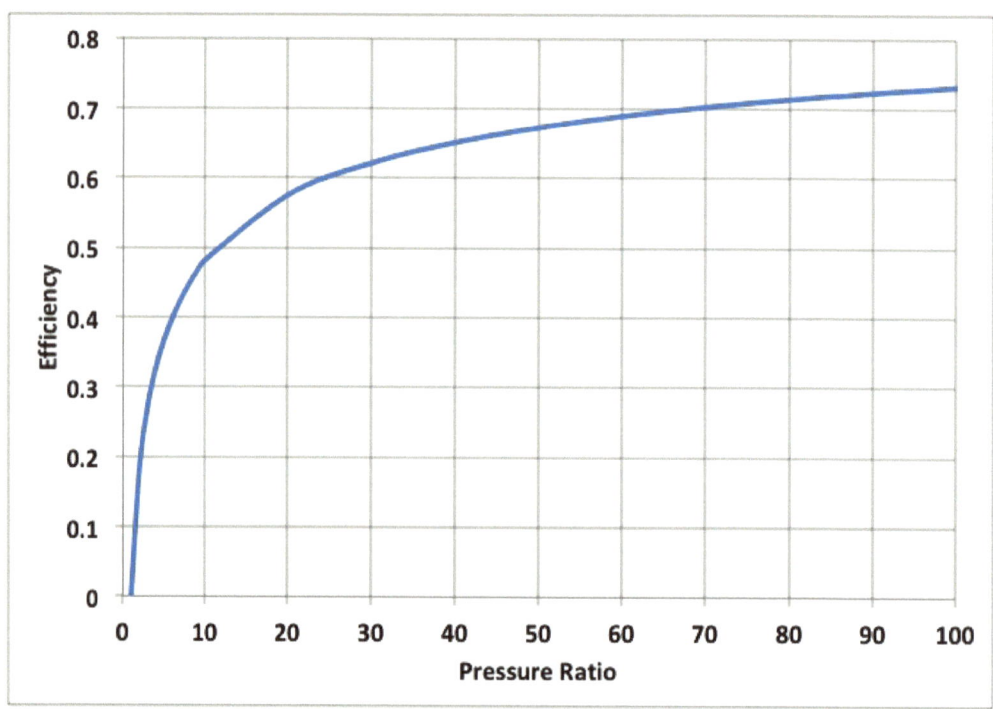

Figure 1.18: First law efficiency as a function of pressure ratio for ideal cold-air standard Brayton cycle.

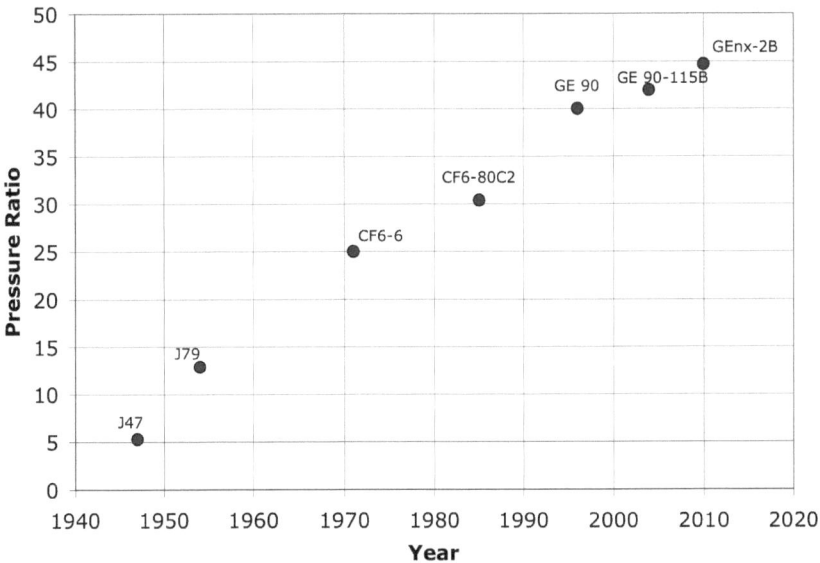

Figure 1.19: Historical trend of compressor pressure ratio in production jet aircraft engines.

Example 1.10
Compute the efficiency and the states of the simple Brayton cycle with a pressure ratio of 20 and heat addition of 1400 kJ/kg.

SOLUTION: Since the compression process is assumed to be isentropic, we can use the isentropic relations for an ideal gas to calculate the temperature at the compressor exit:

$$T_2 = T_1 \left(\frac{P_2}{P_1}\right)^{(k-1)/k} = 300K(20)^{(1.4-1)/1.4} = 706K$$

Assuming a constant pressure combustor, and constant specific heats, the temperature at the outlet of the combustor can be calculated from:

$$T_3 = T_2 + \frac{Q/m}{c_P} = 706K + \frac{1400 kJ/kg}{1.005 kJ/kg \cdot K} = 2099K$$

The turbine expands the exhaust gases out to atmospheric pressure, so the final temperature can be calculated as:

$$T_4 = T_3 \left(\frac{P_4}{P_3}\right)^{(k-1)/k} = 2099K \left(\frac{1}{20}\right)^{(1.4-1)/1.4} = 892K$$

The heat rejected to the environment can be calculated:

$$\frac{Q_{OUT}}{m} = c_p(T_4 - T_1) = 1.005 \frac{kJ}{kg \cdot K}(892K - 300K) = 595 \frac{kJ}{kg}$$

The efficiency can be calculated from an energy balance as:

$$\eta = \frac{1400 kJ/kg - 595 kJ/kg}{1400 kJ/kg} = 0.575 = 57.5\%$$

Table 1.7: Results for Brayton Cycle Example 1.10

State	Pressure (bar)	Temperature (K)
1	1.0	300
2	20.0	706
3	20.0	2099
4	1.0	892

1.7 Stirling Cycle

The Stirling engine was patented by the Rev. Robert Stirling in 1816 and first constructed in 1818. The Stirling cycle models the Stirling Engine, which is a type of *external combustion engine*. In 1843 James Stirling modified one of his brother's engine designs using an existing steam engine and achieved an efficiency of 18%. In the early 1900's Stirling engines were used for powering cooling fans and water pumps. Many variations on the original Stirling design have been built, such as Ericcson and Rider engines, and also some that use solar collectors instead of combustion as the heat source. Figure 1.20 shows the 4 processes of Stirling cycle on a P-V diagram. The displacer piston moves the working fluid back and forth between the hot expansion space and the cold compression space, alternately heating and cooling the working fluid. The power piston is synchronized so that it expands when the working fluid is at its hottest and it compresses the working fluid when the displacer moves the cold working fluid into the working volume. Thus the net work is equal to the expansion work minus the compression work.

Table 1.8: Steps in Stirling Cycle:

Process	Thermodynamic relations	Volume relations
Isothermal Compression	$P_1V_1 = P_2V_2$	$V_2 = V_1$ / Compression ratio
Constant Volume Heat Addition	$Q_{IN} = m\ c_V\ (T_3 - T_2)$	$V_3 = V_2$
Isothermal Expansion	$P_4V_4 = P_3V_3$	$V_4 = V_3$ * Compression ratio
Constant Volume Heat Rejection	$Q_{OUT} = m\ c_V\ (T_4 - T_1)$	$V_4 = V_1$

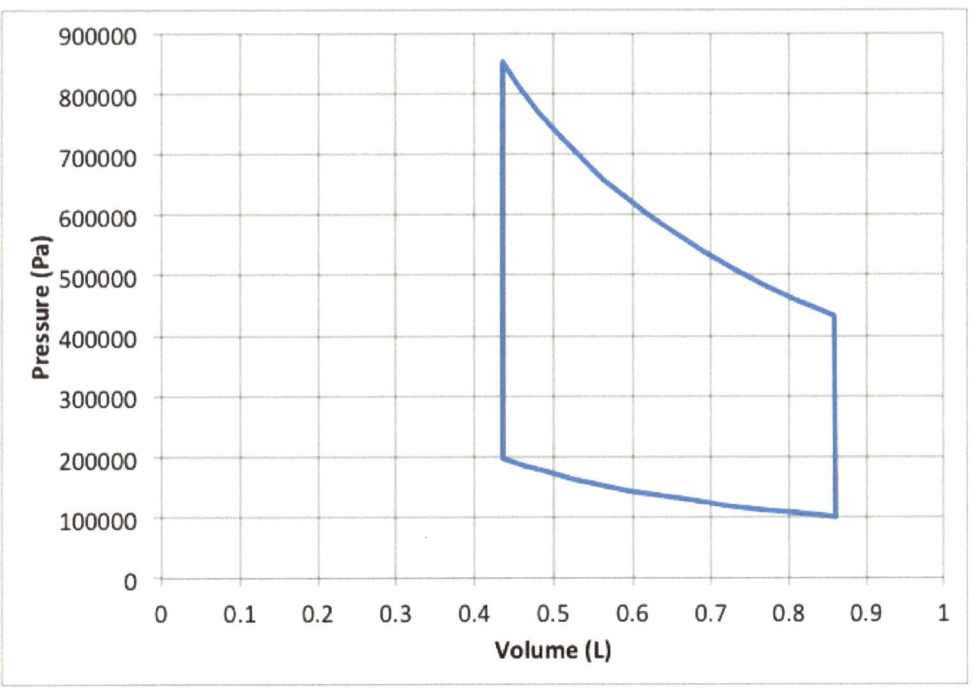

Figure 1.20: P-V diagram of Stirling cycle

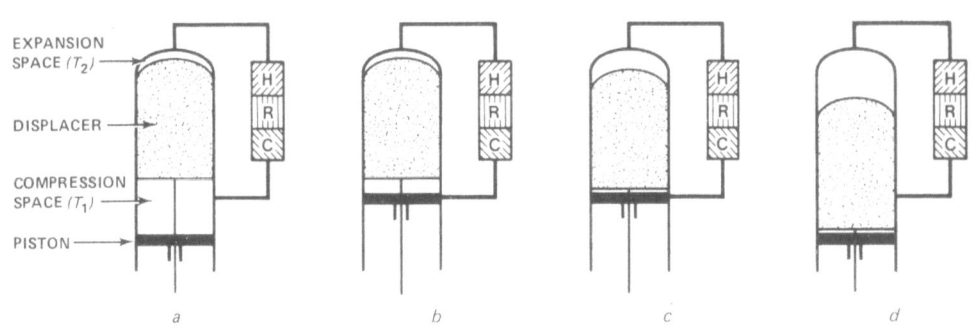

Figure 1.21: Schematic of steps in Stirling cycle. R= regenerator, H = heater, C= cooler. [Deskbook]

The Stirling engine has the potential for very high thermodynamic efficiency, but only when used with a regenerator. A regenerator captures some of the waste heat in process 4-1 and re-uses it as input heat in process 2-3. The best working fluid for a Stirling cycle engine is hydrogen, due to its high thermal conductivity and the lower viscosity, which reduces pumping losses. Helium can also be used [NASA90]. Increasing the pressure of the working gas increases the density and specific power of the engine, so such systems are pressurized.

From time to time the Stirling engine sees renewed interest as a possible engine for automobiles. The potential advantages for a Stirling engine relative to an Otto-cycle-based gasoline engine include low pollutant emissions, low noise levels, long operating life, the theoretical potential for high efficiency, especially at part load, and the ability to use a variety of fuels regardless of octane number. The disadvantages are that cold startup in a Stirling engine is slower than in a gasoline engine, and once hot it also changes speeds more slowly due to its high thermal mass [Deskbook] [Martini83]. A Stirling engine is probably more suitable for a hybrid drivetrain than as the primary powerplant for a passenger vehicle.

1.8 Other Cycles

Two of the limitations of Otto-cycle based gasoline engines are: 1) that the maximum compression ratio is limited by knock, which lowers the maximum possible efficiency of the engine, and 2) that the 4-stroke cycle produces only one power stroke per every two revolutions of the crankshaft. This section details some strategies to overcome these limitations.

Atkinson Cycle

The Atkinson Cycle is a modification of the Otto cycle, in which the expansion stroke is longer than the compression stroke. This can be accomplished by an arrangement of mechanical linkages, such that the expansion and exhaust strokes have the same length, but are longer than the compression and intake strokes. Thermodynamic efficiency increases with higher compression ratio, but gasoline engines are limited in the maximum compression they can obtain due to knocking of the fuel. The Atkinson cycle allows the benefits of a high expansion ratio to be obtained while avoiding the problem of knocking by limiting the compression ratio. Figure 1.22 shows the P-V diagram for an Atkinson Cycle.

While engineers have generally decided that the advantages of the Atkinson Cycle are not worth the penalty of extra weight and complexity, some recent designs have achieved the effect of an Atkinson cycle without the need for additional mechanical linkages. This is accomplished through changes in valve timing from what is standard for normal Otto-cycle engines. Then the intake valve is held open past the end of the intake stroke and during part of the compression stroke, which forces some of the air back into the intake manifold. Thus compression does not effectively start until after the piston is already part of the way up through the compression stroke. The net result is that the compression ratio is smaller than the expansion ratio. The penalty paid relative to a conventional engine is a lower power density, because the full displacement volume of the engine is not filled with air at the intake manifold pressure. In a fully-expanded Atkinson cycle the pressure at the end of the expansion stroke would equal atmospheric pressure.

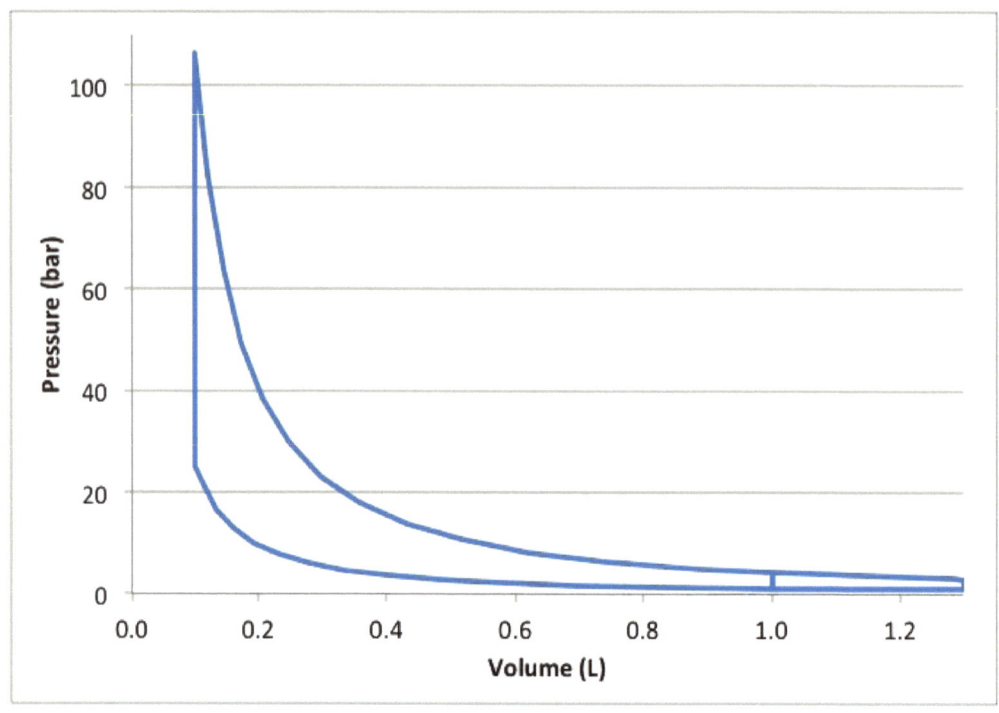

Figure 1.22: P-V diagram of Atkinson cycle.

As a specific example, the Toyota Prius has a geometric compression ratio of 13:1, but because of the late intake valve closing, the effective compression ratio is only about 9.5:1. The loss in power density for the Prius is compensated for by its hybrid electric drivetrain, in which the electric motor can provide the low-end torque the vehicle needs for accelerating from a stop. The Toyota Camry Hybrid hybrid electric sedan has a geometric compression ratio of 12.5:1. For comparison, most gasoline engines on the market now have a compression ratio of around 10:1.

Example 1.11

Calculate the efficiency of a cold-air standard Atkinson Cycle with a compression ratio of 10 and an expansion ratio of 13, with intake at 1 bar and 300 K, and heat input of 1750 kJ/kg. How does this compare to an Otto Cycle with a compression ratio of 10 under comparable conditions?

SOLUTION: Starting with intake conditions of 1 bar and 300 K, the conditions after compression will be:

$$P_2 = P_1(CR)^k = 1bar(10)^{1.4} = 25.1bar$$

$$T_2 = T_1(CR)^{k-1} = 300K(10)^{0.4} = 754K$$

For constant volume combustion with Q/m = 1750 kJ/kg, the temperature after combustion is:

$$T_3 = T_2 + \frac{Q/m}{c_V} = 754K + \frac{1750 kJ/kg}{0.718 kJ/kg \cdot K} = 3191K$$

and the pressure after combustion is calculated from the ideal gas law:

$$P_3 = P_2 \frac{T_3}{T_2} = 25.1bar \frac{3191K}{754K} = 106.2bar$$

The pressure and temperature after the expansion stroke are:

$$P_4 = P_3 \left(\frac{1}{CR}\right)^k = 106.2bar \left(\frac{1}{13}\right)^{1.4} = 2.93bar$$

$$T_4 = T_3 \left(\frac{1}{CR}\right)^{k-1} = 3191K \left(\frac{1}{13}\right)^{0.4} = 1144K$$

The efficiency is calculated using an energy balance where $Q_{OUT} = c_V(T_4-T_1) = 0.718$ kJ/kg-K (1144 K − 300 K) = 606 kJ/kg.

$$\eta = \frac{Q_{IN} - Q_{OUT}}{Q_{IN}} = \frac{1750 kJ/kg - 606 kJ/kg}{1750 kJ/kg} = 0.654 = 65.4\%$$

For comparison, the Otto cycle at these conditions the efficiency would be:

$$\eta = 1 - \frac{1}{CR^{k-1}} = 1 - \frac{1}{10^{0.4}} = 0.602 = 60.2\%$$

Miller Cycle

A Miller Cycle is an Atkinson Cycle with a supercharger added to increase the pressure and density of the air inducted during the intake stroke. This compensates for the lost mass of charge due to leaving the intake valve open during the beginning of compression in an Atkinson engine. The Miller cycle was originally proposed by Ralph Miller in the 1940s.

Two-Stokes and Rotary Engines

A two-stroke engine fires once for every revolution of the crankshaft, compared to a four-stroke engine that has only one combustion event for every two rotations of the crankshaft. Two stroke engines do not have valves, instead air is drawn into the combustion chamber through an intake port, which is uncovered when the piston moves low enough during the intake stroke, and similarly the exhaust is pushed out through an exhaust port, so that the intake and exhaust processes occur simultaneously. A schematic of the process is shown in Figure 1.23. Two stroke engines are commonly used in applications where small, lightweight powerplants are needed, such as chainsaws, leaf blowers, small dirt bikes, and model RC vehicles, though in the last category the two-stroke engines are increasing being replaced by brushless DC motors. Two-stroke engines are generally less efficient that four-stroke engines, with lower compression ratios and relatively inefficient intake and exhaust processes. Two-stroke diesel engines have found use in large marine diesel engines. Since a diesel engine is direct injected, it does not have the problems of short-circuiting the fuel.

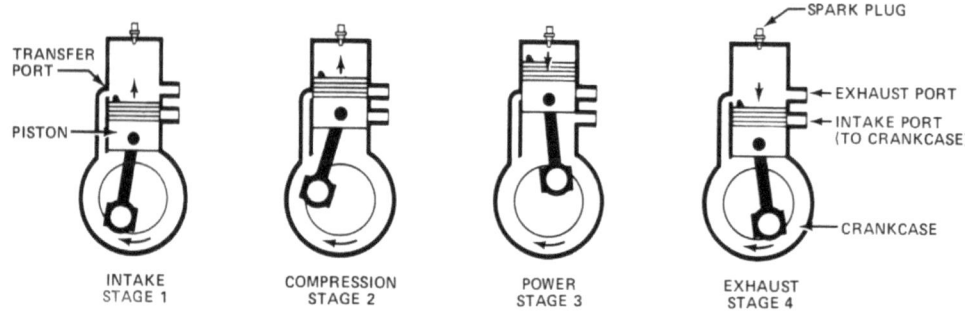

Figure 1.23: Schematic of processes in a 2-stroke engine. [Deskbook]

A rotary engine (also called a Wankel engine) is similar to a two-stroke in that there is one power stroke for every revolution of the crankshaft. As shown in Figure 1.24, there are three separate volumes divided by a single rotor, so that 3 combustion strokes are obtained per rotation of the rotor. The rotors have a trochoidal shape, and move in an epitrochoidal housing. The tips of the rotor provide the sealing to separate the three chambers. The rotor has three firing pulses per revolution of the rotor, or one firing pulse per revolution of the flywheel, which spins at three times the rotor speed. As with the traditional 2-stroke, there are no valves, and intake and exhaust flows are through ports that are opened and closed and the rotor moves.

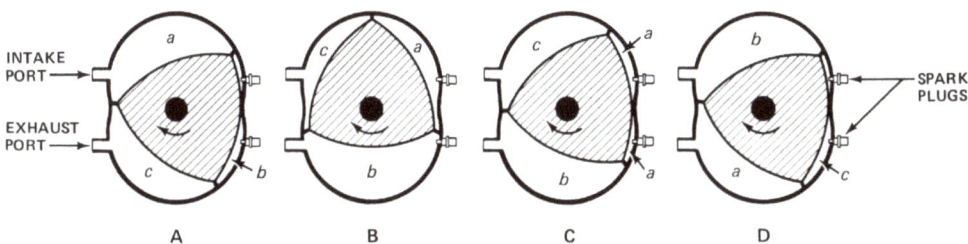

Figure 1.24: Schematic of processes in a rotary (Wankel-type) engine. [Deskbook]

Felix Wankel (1902-1988) patented his first rotary engine design in 1934. Mazda introduced the first production car with a rotary engine in 1967. The RX-2 was the first rotary engine car sold in the United States in 1970. Disadvantages to rotary engines include oil consumption. The RX-7 was introduced in 1978. [Hege01] The Renesis engine in the Mazda RX-8 produces a peak power of 173 kW (232 hp) with only a 1.3 L displacement over 2 rotors, and a 10:1 compression ratio. The Wankel engine has a high power to weight and power to size ratio, but low efficiency compared to standard reciprocating 4-stroke engines [Deskbook]. Advantages of the Wankel are low vibrations (since there is no valvetrain), as well as no reciprocating pistons.

1.9 Fuels

This section briefly discusses some properties of fuels important to internal combustion engines. The energy released in the chemical reaction when a fuel is burned in an engine is converted to mechanical power by the engine. The chemical energy is released when the carbon and hydrogen in the fuel molecule is oxidized. The total enthalpy of a substance is defined as the sum of the chemical enthalpy and the sensible enthalpy. The sensible enthalpy is the energy associated with the thermal state of the material, which is what most first semester thermodynamics courses are concerned with.

$$h_i(T) = h°_{f,i}(T_{ref}) + \Delta h_{s,i}(T-T_{ref}) \qquad (1.33)$$

T_{ref} is usually taken to be 25 °C = 298.15 K and the corresponding P_{ref} is 1 atm = 101,325 Pa. The heat of formation for an element in its naturally occurring state is zero. For example, at STP air is in its naturally occurring state of a gas, so the heat of formations for gaseous O_2 and N_2 at 298 K are 0. Table 1.9 shows the chemical enthalpies for chemical species commonly encountered in the combustion of hydrocarbon fuels.

Table 1.9: Chemical enthalpy of formation for common molecules in combustion systems.

Name	Chemical Formula	Enthalpy of formation at 298 K (kJ/kmol)
Methane (gas)	CH_4	-74,870
Hydrogen (gas)	H_2	0
Oxygen (gas)	O_2	0
Carbon Monoxide	CO	-110,530
Carbon Dioxide	CO_2	-393,520
Water Vapor	H_2O	-241,820
Water (liquid)	H_2O	-285,820

The **heat of combustion** for a chemical reaction is defined as the net change in chemical energy from reactants to products:

$$\Delta H_{comb} = \sum_{Reactants} N_i h^o_{f,i} - \sum_{Products} N_i h^o_{f,i} \qquad (1.34)$$

where N_i is the number of moles of each species in the balanced reaction, based on one mole of fuel.

Example 1.12: Calculate the heat of combustion for methane (CH_4).

SOLUTION: First we must write the balanced chemical reaction for methane combusting with oxygen, which is:

$$CH_4 + 2\,O_2 \rightarrow CO_2 + 2\,H_2O$$

From Table 1.9, for methane the heat of formation is -74,870 kJ/kmol, for oxygen is 0, for carbon dioxide is -393,520 kJ/kmol, and for water vapor is -241,820 kJ/kmol. (The molecular weight of methane is 1*12 + 4*1 = 16 kg/kmol, so on a mass basis the heat of formation is -74,870 kJ/kmol / 16.04 kg/kmol = -4668 kJ/kg.) To calculate the heat of combustion using Eq. 6.2 we write:

$$\Delta H = h^0_{f,CH4} + 2h^0_{f,O2} - h^0_{f,CO2} - 2h^0_{f,H2O}$$

Substituting in the numerical values:

$$\Delta H = -74{,}870 \text{ kJ} + 0 \text{ kJ} - (-393{,}520 \text{ kJ}) - 2 \times (-241{,}820 \text{ kJ}) = 802{,}290 \text{ kJ}$$

Thus, the heat of combustion is $\Delta H = 802{,}290$ kJ per kmol of CH_4. Since heating values are more commonly reported on a mass basis, we can convert this using the molecular weight of methane, M = 16.04 kg/kmol.

$$\Delta H = 802{,}290 \text{ kJ/kmol} / 16 \text{ kg/kmol} = 50{,}018 \text{ kJ/kg}$$

The value of the heat of combustion calculated in Example 1.12 is referred to as the **Lower Heating Value** (LHV). The LHV is applicable when the water in the combustion products is in the gaseous state. The enthalpy of formation for water vapor is -241,820 kJ/kmol, while for liquid water it is -285,820 kJ/kmol. If the water in the products can be condensed to the liquid state, then more heat can be released, and the resulting heat of combustion is referred to as the **Higher Heating Value** (HHV). So for the specific example of methane, if we replace the water vapor in the products with liquid water, then the heat of combustion is:

$$\Delta H = -78{,}870 \text{ kJ} + 0 \text{ kJ} - (-393{,}520 \text{ kJ}) - 2 \times (-285{,}820 \text{ kJ}) = 886{,}290 \text{ kJ}$$

Which on a per mass basis is $\Delta H = 886{,}290$ kJ/kmol / 16.04 kg/kmol = 55,255 kJ/kg. Thus for methane the higher heating value is 10.5% larger than the lower heating value. See Table 1.10 for other heats of combustion for different fuels.

The **stoichiometric air to fuel ratio** is the ratio of the mass of air to the mass of fuel when there is just enough oxygen to complete combust all the carbon to CO_2 and the hydrogen to H_2O for hydrocarbon fuels. Standard dry atmospheric air is comprised of 78.08% nitrogen, 20.95% oxygen, 0.93% argon, 0.03% carbon dioxide, and some trace gases [USATM76]. From a combustion point of view, only the oxygen participates in the combustion reaction, so we can simplify by assuming air is 21% oxygen and 79% nitrogen on a molar basis. Thus for every one mole of oxygen we need, we must also bring in 0.79/0.21 = 3.76 moles of nitrogen. For most of the hydrocarbon fuels this ratio is around 15:1.

Example 1.13: Find the stoichiometric air to fuel ratio for propane (C_3H_8).

SOLUTION: First we need to write the balanced combustion reaction for propane, based on 1 kmol of fuel:

$$C_3H_8 + \_\_ (O_2 + 3.76\ N_2) \rightarrow \_\_ CO_2 + \_\_ H_2O + \_\_ N_2$$

Since there are 3 carbon atoms in the fuel, there must be three CO_2 molecules in the products, and since there are 8 hydrogen atoms in the fuel, there must be 4 H_2O molecules in the products.

$$C_3H_8 + \_\_ (O_2 + 3.76\ N_2) \rightarrow 3\ CO_2 + 4\ H_2O + \_\_ N_2$$

Then there must be 3 + 4/2 = 5 oxygen molecules in the reactants to balance the reaction, with 5*3.76 = 18.8 nitrogen molecules carried through, giving a final overall combustion reaction for propane in air of:

$$C_3H_8 + 5.0\ (O_2 + 3.76\ N_2) \rightarrow 3\ CO_2 + 4\ H_2O + 18.8\ N_2$$

Note: 3.76/(1.0+3.76) = 0.79 is the mole fraction of N_2 in air. The mass of fuel is m_f = 3*12 + 8*1 = 44 kg, and the mass of air is m_a = 5*(32 + 3.76*28) = 686.4 kg. Thus the overall air-fuel ratio is: A/F = 686.4 kg/44 kg = 15.6.

Another popular measure of the stoichiometry is the **equivalence ratio**, f. Since the stoichiometric air to fuel ratio varies between different fuels, the equivalence ratio was defined to be equal to 1 at stoichiometric for all fuels. This is accomplished by normalizing the actual air to fuel ratio for a combustion device by the theoretical air to fuel ratio for that fuel under stoichiometric conditions. The mathematical definition of the equivalence ratio is:

$$f = (A/F)_{stoic} / (A/F)_{actual} \qquad (1.35)$$

where A/F is the air to fuel ratio on a mass basis. When $f < 1$, the system is referred to as lean, f = 1 is stoichiometric, and $f > 1$ is rich. How do you write the reaction when $f \neq 1$? There are two choices. You can write:

or

$$f C_xH_y + (x + y/4)(O_2 + 3.76 N_2) \rightarrow \text{products}$$

$$C_xH_y + (1/f)(x + y/4)(O_2 + 3.76 N_2) \rightarrow \text{products} \quad (1.36)$$

The composition of the products will depend on whether the system is rich or lean. For $f < 1$ (lean) we expect all the fuel to be able to burn completely to CO_2 and H_2O, but also expect there to be some extra leftover O_2 in the products. Thus the balanced chemical reaction for any hydrocarbon fuel can be written as:

$$C_xH_y + (x + y/4)(O_2 + 3.76 N_2) \rightarrow x\,CO_2 + y/2\,H_2O + 3.76(x + y/4) N_2$$

Example 1.14

Calculate the stoichiometric air to fuel ratio and the lower heating value for gaseous ethanol (C_2H_5OH).

SOLUTION: The balanced chemical reaction is:

$$C_2H_5OH + 3(O_2 + 3.76 N_2) \rightarrow 2\,CO_2 + 3\,H_2O + 11.28\,N_2$$

Note the oxygen in the fuel must be accounted for in the oxygen balance. The air to fuel ratio on a mass basis is: $3(32 + 3.76(28))/(2*12+6*1+16) = 411.84/46 = 8.95$. Note that for most hydrocarbon fuels the stoichiometric air to fuel ratio is around 15:1. To calculate the lower heating value, we need the enthalpy of formation of ethanol, which is $-235,300$ kJ. Using the enthalpies of formation of the products in Table 6.1, the change in enthalpy is $-235,300$ kJ $- (2*(-393,520$ kJ$) + 3*(-241,820$ kJ$)) = 1,277,200$ kJ per kmol of ethanol. On a per mass basis, the lower heating value is $(1,277,200$ kJ/kmol$)/(46$ kg/kmol$) = 27,800$ kJ/kg.

Advantages of ethanol include that it generally results in better emissions from engines, and has higher resistance to knock than regular gasoline. One disadvantage is that it has a lower energy content, in kJ/kg, resulting in lower vehicle mileage and range. E10 fuel is a mixture of 10% ethanol and 90% gasoline. A car will typically get 3-4% worse fuel economy (mpg) with E10 than with straight gasoline [NREL09]. E85 fuel is a mixture of 85% ethanol and 15% gasoline, which is used in some flexible fuel vehicles. A car running on E85 will typically obtain 25-30% less fuel economy (mpg) than gasoline [West07]. Ethanol has an octane number of about 105, and for E85 it is 101.

Table 1.10: Properties of common combustion fuels. Data from [DOE94], [NACA56], [NIST08]. Heating value in megajoules. 1 MJ = 1000 kJ.

Fuel	Chemical Formula	Molecular Weight (kg/kmol)	Stoichiometric A/F ratio	Lower Hearing Value (MJ/kg)
Hydrogen	H_2	2.02	34.3	120.0
Methane	CH_4	16.04	17.2	50.0
Propane	C_3H_8	44.1	15.7	46.3
Methanol	CH_3OH	32.04	6.5	20.1
Ethanol	C_2H_5OH	46.07	9.0	27.0
Gasoline	mixture	~102	14.7	42.4
Diesel Fuel	mixture	~200	14.7	42.6

1.10 Advances in Engine Technology

While it is known that engine efficiency can be increased by increasing the compression ratio, in premixed-charge gasoline engines the compression ratio is limited by premature detonation, or knocking, which in turn relates to the octane number of the fuel. (Fuel energy content is not affected by the octane number). Even in diesel engines, where the fuel is directly injected into the cylinder and not premixed, the compression ratio is limited by the mechanical strength of the connecting rod, as well as blow-by of high pressure gases past the piston rings. Car manufacturers are continuously implementing new technologies to improve vehicle drivetrain efficiency. Among the recent innovations becoming more common are gasoline direction injection (GDI) engines and continuously-variable transmissions (CVT). Figure 1.25 shows the market penetration of some of these recent technological innovations.

Further, there is an increasing number of hybrid gasoline-electric powertrain cars and pure electric vehicles on the market. The US EPA's top ten mileage lists (http://www.fueleconomy.gov/feg/topten.jsp) are dominated by these cars. Table 1.11 shows a selection of currently available (as of 2018) electric cars with important specifications.

Table 1.11: Examples of electric vehilces available in the U.S. Battery energy storage capacity in kilowatt-hours. 1 kW-hr = 3600 kJ.

Vehicle	Cost (USD)	Range (miles)	Motor (hp)	Battery (kW-hr)
Mitsubishi i-Miev	$23,000	62	66	16
Ford Focus Electric	$29,000	115	143	34
Hyundai Ioniq	$30,000	124	120	28
Volkswagen e-Golf	$30,000	125	134	36
Nissan Leaf	$31,000	151	147	24
Tesla Model 3	$50,000	220	265	50
Tesla Model S	$76,000	335	382	100

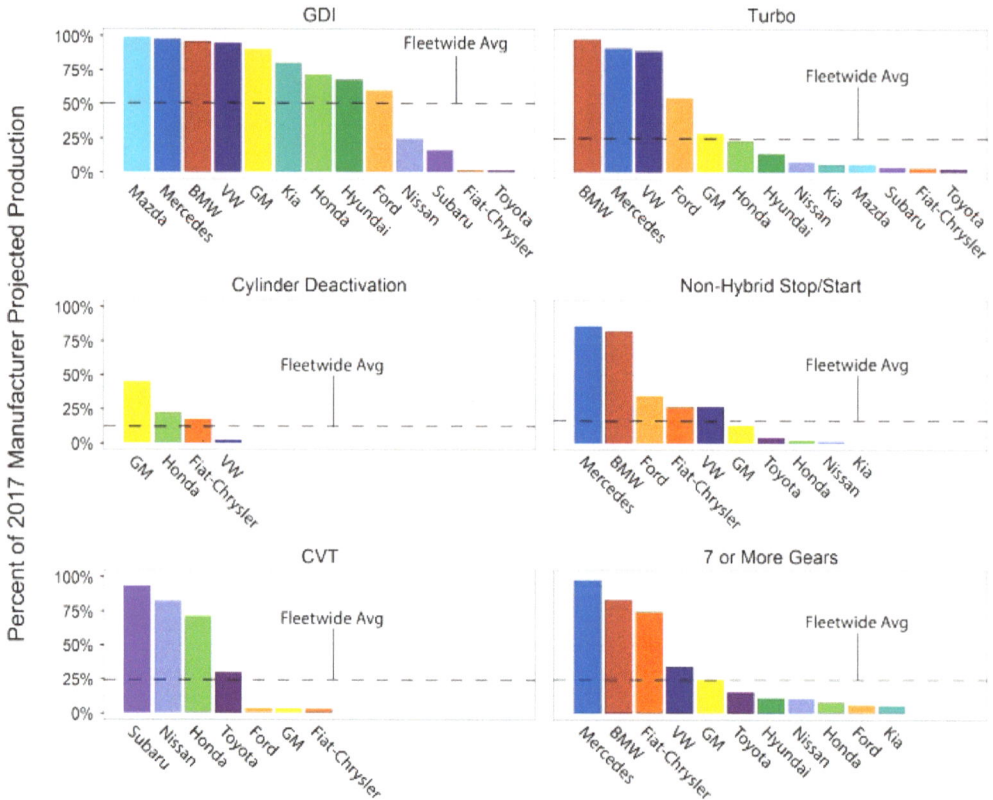

Figure 1.25: Adoption of fuel-efficiency improving technologies in production cars in the U.S. Source: https://www.epa.gov/fuel-economy-trends/highlights-co2-and-fuel-economy-trends

Useful Websites

- A free alternative to MATLAB:
    - Octave - http://www.gnu.org/software/octave/
- MATLAB Tutorials can be found at:
    - http://www.engin.umich.edu/group/ctm/basic/basic.html
    - http://web.cecs.pdx.edu/~gerry/MATLAB/
    - http://www.math.mtu.edu/~msgocken/intro/intro.html
- http://www.animatedengines.com/
- http://www.kohlerengines.com/difference/howengineswork.htm
- http://world.honda.com/automobile-technology/index.html
- http://world.honda.com/automobile-technology/VTEC/
- http://www.compcams.com/Base/Images/Technical/800-615-ValveTimingIllustration-002.gif
- http://www.kewengineering.co.uk/Auto_oils/oil_viscosity_explained.htm
- https://www.engineeringtoolbox.com/dynamic-absolute-kinematic-viscosity-d_412.html
- https://www.fueleconomy.gov/feg/topten.jsp
- https://auto.howstuffworks.com/diesel.htm
- https://auto.howstuffworks.com/engine.htm
- https://science.howstuffworks.com/transport/engines-equipment/two-stroke.htm
- https://auto.howstuffworks.com/turbo.htm
- https://auto.howstuffworks.com/cooling-system.htm
- https://auto.howstuffworks.com/fuel-injection.htm
- https://auto.howstuffworks.com/rotary-engine.htm
- https://auto.howstuffworks.com/stirling-engine.htm

THERMODYNAMICS EQUATION SHEET

First Law: $\Delta E = Q - W$

Ideal Gas Law: $PV = NR_u T$ $m = N*M$ $v = V/m$ $\rho = PM/R_u T$

$R_u = 8314$ J/kmol-K $R = R_u / M$

Specific Heats: $c_P = dh/dT$ $c_V = du/dT$ $R = c_P - c_V$ $k = c_P / c_V$

Properties of air @ STP: $k = 1.4$ $M = 29$ kg/kmol $R = 287$ J/kg-K

$c_P = 1005$ J/kg-K $c_V = 718$ J/kg-K

Boundary Work: $W = \int P\, dV$ Isothermal process of ideal gas: $PV = C$

Isentropic process of ideal gas: $PV^k = C$

Conservation of mass: $\dfrac{dm}{dt} = \sum_{in} \dot{m} - \sum_{out} \dot{m}$ $\dot{m} = \rho A V$

Isentropic process of an ideal gas: $\dfrac{P_2}{P_1} = \left(\dfrac{V_1}{V_2}\right)^k$ $\dfrac{T_2}{T_1} = \left(\dfrac{V_1}{V_2}\right)^{k-1}$ $\dfrac{T_2}{T_1} = \left(\dfrac{P_2}{P_1}\right)^{\frac{k-1}{k}}$

Carnot cycles: $Q_H / Q_L = T_H / T_L$ $\eta = 1 - T_L / T_H$ for Carnot power cycle

$\eta_{Otto} = 1 - \dfrac{1}{CR^{k-1}}$ $\eta_{Brayton} = 1 - \dfrac{1}{PR^{(k-1)/k}}$

Combustion: For air, for every 1 kmol of O_2 there is 3.76 kmol of N_2

Lower heating values (kJ/kg) at reference state of 298 K and 1 atm (101.3 kPa)

	CH_4	C_2H_6	C_3H_8	C_4H_{10}	C_5H_{12}	C_8H_{18}
$\Delta h°_f$ (kJ/kmol)	-74,830	-84,670	-103,850	-124,730	-146,440	-208,450
LHV (kJ/kg)	50,020	47,490	46,360	45,740	45,360	44,790

Atomic weights: H – 1, C – 12, N – 14, O – 16

Conversion factors:

1 hp = 550 ft-lbf/s   4.45 N = 1 lbf   1 m = 3.28 ft

1 lbf = 32.2 lbm-ft/s^2 1 hp = 746 W   1 m/s = 2.24 mph   1 mile = 5280 ft

1 Gal = 231 in^3 = 3.785 L   1000 L = 1 m^3 2.54 cm = 1 in   1 kg = 2.2 lbm

1 atm = 14.7 psi = 101,325 Pa   1 bar = 100,000 Pa = 14.5 psi   1 ft-lbf = 1.357 N-m

1 mi = 5280 ft = 1.609 km = 1609 m   [°F] = 9/5 [°C] + 32   [°C] = ([°F] - 32)*5/9

32 mpg = 13.6 km/L = 7.35 L/100 km

Summary Comparison of Engine Cycles

Otto Cycle

Process	Thermodynamic relations	
Isentropic Compression	$P_1V_1^k = P_2V_2^k$	$V_2 = V_1$ / Compression ratio
Constant Volume Heat Addition	$Q_{IN} = m\, c_V\, (T_3 - T_2)$	$V_3 = V_2$
Isentropic Expansion	$P_4V_4^k = P_3V_3^k$	$V_4 = V_3$ * Compression ratio
Constant Volume Heat Rejection	$Q_{OUT} = m\, c_V\, (T_4 - T_1)$	$V_4 = V_1$

Diesel Cycle

Process	Thermodynamic relations	
Isentropic Compression	$P_1V_1^k = P_2V_2^k$	$V_2 = V_1$ / Compression ratio
Constant Pressure Heat Addition	$Q_{IN} = m\, c_P\, (T_3 - T_2)$	$P_3 = P_2$
Isentropic Expansion	$P_4V_4^k = P_3V_3^k$	$V_3 = V_2$ * Cutoff ratio
Constant Volume Heat Rejection	$Q_{OUT} = m\, c_V\, (T_4 - T_1)$	$V_4 = V_1$

Dual Cycle

Process	Thermodynamic relations	
Isentropic Compression	$P_1V_1^k = P_2V_2^k$	$V_2 = V_1$ / Compression ratio
Constant Volume Heat Addition	$Q_{IN} = m\, c_V\, (T_3 - T_2)$	$V_3 = V_2$
Constant Pressure Heat Addition	$Q_{IN} = m\, c_P\, (T_4 - T_3)$	$P_4 = P_3$
Isentropic Expansion	$P_5V_5^k = P_4V_4^k$	$V_4 = V_3$ * Cutoff ratio
Constant Volume Heat Rejection	$Q_{OUT} = m\, c_V\, (T_5 - T_1)$	$V_5 = V_1$

Stirling Cycle

Process	Thermodynamic relations	
Isothermal Compression	$P_1V_1 = P_2V_2$	$V_2 = V_1$ / Compression ratio
Constant Volume Heat Addition	$Q_{IN} = m\, c_V\, (T_3 - T_2)$	$V_3 = V_2$
Isothermal Expansion	$P_4V_4 = P_3V_3$	$V_4 = V_3$ * Compression ratio
Constant Volume Heat Rejection	$Q_{OUT} = m\, c_V\, (T_4 - T_1)$	$V_4 = V_1$

Brayton Cycle

Process	Thermodynamic relations	
Isentropic Compression	$T_2 = T_1*(P_2/P_1)^{(k-1)/k}$	$P_2 = P_1 *$ Compressor P ratio
Constant Pressure Heat Addition	$Q_{IN} = m\ c_P\ (T_3 - T_2)$	$P_3 = P_2$
Isentropic Expansion	$T_4 = T_3*(P_4/P_3)^{(k-1)/k}$	$P_4 = P_3 /$ Compressor P ratio
Constant Pressure Heat Rejection	$Q_{OUT} = m\ c_P\ (T_4 - T_1)$	$P_4 = P_1$

References

[NACA33] Fuel Vaporization and Its Effect on Combustion in a High-Speed Compression-Ignition Engine. NACA Report 435. A. Rothrock and C. Waldron. 1933.

[Eichelberg39] Some New Investigations on Old Combustion Engine Problems. Engineering. Vol. 149. G. Eichelberg. 1939.

[NACA56] R. Hibbard. Evaluation of Liquefied Hydrocarbon Gases as Turbojet Fuels. NACA-RM-E56I21. 1956.

[Annand63] Heat Transfer in the Cylinders of Reciprocating Internal Combustion Engines. Proceedings of the Institution of Mechanical Engineers, Vol. 177. W. Annand. 1963.

[USATM76] U.S. Standard Atmosphere. NOAA. 1976.

[Deskbook] S. Gladstaone. Energy Deskbook. U.S. Dept. of Energy. 1982.

[Martini83] W. Martini, Stirling Engine Design Manual, DOE/NASA/3194-1, 1983.

[Heywood88] Internal Combustion Engine Fundamentals. McGraw-Hill. J. Heywood. 1988.

[NASA90] Stirling Commercialization Study. NASA-TM-111664. 1990.

[DOE94] Alternatives to Traditional Transportation Fuels. DOE/EIA-0585/0. 1994.

[Caton00] Comparisons of Instructional and Complete Versions of Thermodynamic Engine Cycle Simulations for Spark-Ignition Engines. J. Caton. *International Journal of Mechanical Engineering Education*, Vol. 29, pp. 283-306, 2000.

[Hege01] The Wankel Rotary Engine. J. Hege. McFarland & Co. 2001.

[West07] B. West, A. Lopez, T. Theiss, R. Graves, J. Storey, and S. Lewis. Fuel Economy and Emissions of the Ethanol-Optimized Saab 9-5. SAE Technical Paper 2007-01-3994. 2007.

[NIST08] NIST Chemistry WebBook, NIST Standard Reference Database Number 69, Eds. P.J. Linstrom and W.G. Mallard, National Institute of Standards and Technology.

[Mittal09] V. Mittal and J. Heywood. The Shift in Relevance of Fuel RON and MON to Knock Onset in Modern SI Engines Over the Last 70 Years. SAE Paper 2009-01-2622.

[Aneja09] Aneja, R., & Kayes, D. (2009). Reduction of heavy-duty fuel consumption and CO_2 generations: What the industry does and what government can do. Presented at the Directions in Engine-Efficiency and Emissions Research Conference (DEER), August 5, 2009.

[NREL09] Effects of Intermediate Ethanol Blends on Legacy Vehicles and Small Non-Road Engines. NREL/TP-540-43543. 2009.

[Databook] S. Davis, S. Diegel, and R. Boundy. Transportation Energy Data Book, 30th edition. U.S. Dept. of Energy. ORNL-6986. 2011. http://cta.ornl.gov/data

MATLAB CODES

```
clear all
% this code computes an engine cycle simulation including heat
% transfer losses, and finite rate of combustion
% all units in m-k-s-K
CR = 12;
bore = 0.1;
stroke = 0.1;
rpm = 1800;
omega = rpm*2*pi/60;
sp = 2*stroke*omega;
rat = 1.5;
k = 1.4;
lambda = 0.15;
M = 29;
R = 8314;
P0 = 100000;
T0 = 300;
Twall = 300;
work = 0;
% set displacement volume to 1 L
% Vdisp = 0.001;
Vdisp = stroke*pi*bore^2/4;
rho = P0*M/(R*T0);
% calculating mass of fuel for near stoichiometric condition
m = rho * Vdisp/(1-1/CR);
% qchem = 1750*1000*m;
mf = m/16;
% heat release for fuel LHV = 40,000 kJ/kg - units of J
qchem = mf * 40000 * 1000;
% parameters for finite time of heat release
tstart = -10.0;
tcomb = 80.0;
% convert time units to radians from degrees
Qrate = qchem/tcomb*180/pi;
% specify dt in degrees
dt = 0.05;
dqdt = 0;
P(1)= P0;
T(1)= T0;
% define TDC to be 0 CAD;
for i=1:360/dt+1;
    theta(i)= -180+(i-1)*dt;
    a = theta(i);
    V(i)=Vdisp/(CR-1)+0.5*Vdisp*(rat+1-cosd(a)-sqrt(rat^2-(sind(a))^2));
```

```
        dvdt(i) = 0.5*Vdisp*sind(a)*(1+cosd(a)/sqrt(rat^2-
(sind(a))^2));
end
% main program loop
for i=1:360/dt;
% instantaneous heat release
%    if theta(i) == 0
%         dqdt = qchem/(dt*pi/180);
%    else
%         dqdt = 0;
%    end
% combustion spread out over finite time
    if (theta(i) > tstart) && (theta(i) < tstart+tcomb)
         dqdt = Qrate;
    else
         dqdt = 0.0;
    end
% heat loss calculation
    h = lambda/bore*0.5*(rho*sp*bore/7.0E-05)^0.7;
    A=pi/2*bore^2+pi*bore*stroke/2*(rat+1-cosd(a)+sqrt(rat^2-
(sind(a))^2));
    dqht = 0.001*h*A*(T(i)-Twall);
%    dqht = 0;
    dqdt = dqdt - dqht;
% turn off combustion for a motoring trace
%    dqdt = 0;
    dpdt = (k-1)/V(i)*dqdt-k*(P(i)/V(i))*dvdt(i);
% convert dt to radians
    P(i+1)=P(i)+dpdt*dt*pi/180;
    T(i+1)=P(i+1)*M*V(i+1)/(R*m);
    work = work + P(i)*(V(i+1)-V(i));
end
% calculate efficiency
eta = work / qchem
% create a P-V diagram
figure
hold on
plot(V,P)
xlabel('volume (m^3)')
ylabel('Pressure (Pa)')
% output data to a text file
fID = fopen('press.txt','w');
for i=1:360/dt;
  fprintf(fID,'%f %f %f \n',theta(i),V(i)*1000,P(i))
end
fclose(fID);
```

```
% code to compute the intake stroke of an engine
clear all
% computing intake stroke and volumetric efficiency
% all units in m-k-s-K
CR = 10;
rat = 1.5;
M = 29;
R = 8314;
P0 = 100000;
T0 = 300;
% engine speed
w = 1667*2*pi/60
avalve = 0.000125;
cd = 0.75;
% set displacement volume to 1 L
Vdisp = 0.001;
rho = P0*M/(R*T0);
mtheo = rho * Vdisp/(1-1/CR)
% specify dt in degrees
dt = 1.0;
dqdt = 0;
P(1)= P0;
% define TDC to be 0 CAD;
for i=1:180/dt+1;
    theta(i)=(i-1)*dt;
    a = theta(i);
    V(i)=Vdisp/(CR-1)+0.5*Vdisp*(rat+1-cosd(a)-sqrt(rat^2-(sind(a))^2));
    dvdt(i) = 0.5*Vdisp*sind(a)*(1+cosd(a)/sqrt(rat^2-(sind(a))^2));
end
m(1) = rho*V(1);
for i=1:180/dt;
    mdot(i) = cd*rho*avalve*sqrt(2*(P0-P(i))/rho);
    m(i+1) = m(i)+mdot(i)*dt*pi/180/w;
    P(i+1) = P(i)*V(i)/V(i+1)*m(i+1)/m(i);
%    dpdt = R*T0/(M*V(i))*(mdot(i)/w-m(i)*dvdt(i)/V(i));
%    P(i+1)=P(i)+dpdt*dt*pi/180;
end
% final mass
mfinal = m(i+1)
figure
hold on
grid on
plot(V,P)
xlabel('volume (m^3)')
ylabel('Pressure (Pa)')
```

www.ingramcontent.com/pod-product-compliance
Lightning Source LLC
Chambersburg PA
CBHW040327220526
45473CB00009B/2587